BHOPAL

Ballinger Series in

BUSINESS IN A GLOBAL ENVIRONMENT

S. Prakash Sethi, Series Editor

Center for Management
Baruch College
The City University of New York

BHOPAL

ANATOMY OF A CRISIS

PAUL SHRIVASTAVA

BALLINGER PUBLISHING COMPANY
Cambridge, Massachusetts
A Subsidiary of Harper & Row, Publishers, Inc.

International Standard Book Number: 0-88730-084-7

Library of Congress Catalog Card Number: 86-26575

Printed in the United States of America

Library of Congress Cataloging-in-Publication Data

Shrivastava, Paul.
  Bhopal : anatomy of a crisis.

  (Ballinger series in business in a global environment)
  Bibliography: p.
  Includes index.
  1. Bhopal Union Carbide Plant Disaster, Bhopal, India, 1984. 2. Union Carbide Ltd. (India) 3. Union Carbide Corporation. 4. Pesticides industry—India—Bhopal—Accidents. 5. Methyl isocyanate—Toxicology—India—Bhopal. I. Title. II. Series.
  HD7269.C45215268   1987   363.1'79   86-26575

  ISBN 0-88730-084-7

To Michelle in New York
and
To Nevita in Bhopal

Contents

List of Figures

List of Tables

Acknowledgments

In today's complex world of social science research, no book is ever written without direct contributions from many individuals and organizations. I have benefitted from the generosity of so many contributors that it is impossible to acknowledge them all here.

For the intellectual environment and the material support that gave me courage to do this study, I am grateful to the Graduate School of Business Administration, New York University, particularly to the Management and Organizational Behavior Area. Dean Richard West and Professor William Guth supported my efforts throughout the two years of this study.

Several people associated with the Industrial Crisis Institute, Inc. directly contributed to the book. They helped me with the original design of the study, discussed many intellectual and political aspects of the project, and critiqued parts of the book. These include Drs. S. Chandrasekhar, Vasant Dhar, David Ecker, Martin Lee, Ian Mitroff, and Charles Perrow. I also thank David Rogers for his comments on an earlier draft of the book.

Data for the study came from over 200 interviews (many of them confidential) with the people of Bhopal, volunteer social work organizations, officials of the government of India and Madhya Pradesh, and managers of Union Carbide Corporation in India and the United States. I cannot name them all in this limited space. Moreover, many of them wish to remain anonymous. I thank them for their openness in discussing sensitive issues and for taking risks in providing data for this study.

Editors are to books what maintenance crews are to chemical plants. Silently and behind the scenes they work hard to keep the system running. This book was fortunate to have three of the most professional editors in the publishing business—Marjorie Rich-man, Barbara Roth, and William Fulton. I thank them for their patience, good sense, and careful editing.

Despite the help of all these people, errors and misconceptions that remain are entirely my responsibility.

Preface

The Bhopal tragedy—the worst industrial crisis in history, triggered two years ago—continues. Understanding the causes and consequences of this crisis, and learning lessons for the future, is the least we can do to commemorate those whose lives were unalterably changed by it. However, an event of such vast social and technological scope and deep human significance defies explanation based on traditional concepts and theories. To deal with its complexity and scope we need new concepts. This book develops such concepts and makes recommendations for resolving the practical, real-world problems associated with crises. It is as much about Bhopal as it is about what I have termed "industrial crises"—crises, triggered by industrial activities, that cause extensive damage to human life and the natural and social environments.

As we approach the end of this century, we will see the proliferation of new and complex technologies in the industrialized world. Simultaneously, more hazardous technologies will enter communities in developing countries. As a consequence, industrial crises are likely to become more frequent and larger in scope. The Chernobyl nuclear plant accident, the *Challenger* shuttle explosion, and the Bhopal disaster are grim reminders of the fact that these crises are universal. They can occur in the most advanced countries and originate in the most sophisticated organizations. They are compelling arguments for the urgency of developing long-term solutions to industrial crisis problems.

This book is intended to initiate a larger program of research in industrial crises. It lays down the basic concepts needed to understand this new problem facing corporations and society and provides a framework for studying industrial crises. The bulk of the book addresses what happened in Bhopal, the consequences of the accident, and the crisis management responses by the company, the government, and the victims and volunteer organizations. Using empirical data from the Bhopal case, it seeks to understand the crisis from the perspectives of the three principal stakeholders— Union Carbide Corporation, the government of India, and the victims. Finally, it articulates the preventive and coping actions that corporations, governments, and communities must take to deal with industrial crises.

The thrust of the book is to make two arguments, one about causes and the other about consequences of industrial crises. Crises occur most commonly in situations where complex technologies are embedded in communities that do not possess the infrastructure to support them. The technological and industrial events that trigger crises are caused by a complex interaction of human, organizational, and technological factors. These events, in turn, interact with economic, social, and political forces to create crisis. Typically, responses to crises are fragmented and rife with conflict. Moreover, they address only symptoms. Thus the fundamental causes of crises are not eliminated because they are often not known with certainty. As a result, the potential for crisis is a constant threat to industrial societies.

There is nothing natural about industrial crises. Nor are they accidental. Their causes are systemic and controlled by social agents. Such crises can be eliminated through judicious choices by corporations, governments, and communities. Technological, organizational, and social policy solutions to reduce the likelihood of crises do exist, and are presented in this book. Some of these solutions are adopted voluntarily by corporations and governments. Other solutions can be implemented consensually by stakeholders. The most difficult and expensive solutions, unfortunately, must be imposed on each stakeholder by the others. This is the political challenge of managing technology.

To meet this challenge, explanations provided here must be understood by decisionmakers who make choices of technology and industrial policies, by community members who bear the risks imposed by industrial products and production systems, and by researchers and educators who influence our thinking about these crucial issues. Therefore, the book is written for a diverse audience. I have tried to make the complex ideas readable without sacrificing intellectual or technical accuracy. Where ever I have failed to do so, I seek the reader's indulgence.

Paul Shrivastava

CHAPTER 1

Crisis in Bhopal

At about 12:40 A.M. on December 3, 1984, Suman Dey looked at the gauges on the control panel in total disbelief. Dey was the control-room operator at the Union Carbide pesticide plant in Bhopal, India, and what he saw was so far out of the ordinary that it terrified him. Inside a storage tank containing the dangerous chemical methyl isocyanate gas (MIC), which was supposed to be refrigerated, the temperature had risen to 77° F. Pressure in the tank, which ordinarily ranged between 2 and 25 pounds per square inch (psi), had risen to 55 psi.

Bewildered by the readings, Dey ran to the storage tank area to investigate the problem. He heard a loud rumbling sound and saw a plume of gas gushing out of the stack in front of him. Dey, along with the MIC supervisor on duty, Shakil Qureshi, and several operators, attempted to control the gas leak by turning on safety devices. Together they tried switching on the refrigeration system to cool the storage tank. They started the scrubber through which the gases were passing and sprayed water on escaping gases, hoping to neutralize them. When all these efforts failed, they fled the plant in panic.

Across the street in a slum hut, a twenty-eight-year-old woman named Ganga Bai was awakened by incoherent shouting. She felt a burning sensation in her eyes and rubbed them, hoping to soothe them. Outside, she saw terror-stricken neighbors running through the narrow gullies between the huts, shouting single words: "Run!" "Gas!" "Death!" She woke up her husband, picked up her two-year-old daughter, and ran out of the hut.

Ganga Bai bypassed the crowd by running on the muddy ledge between a row of huts and the main road. As she ran, she saw death in its most bizarre forms. People were choking and gasping for breath. Some fell as they ran, and some lay on the roadside, vomiting and defecating. Others, too weak to run, tried to clutch onto people passing them in the hope of being carried forward.

After running for several miles, Ganga Bai stopped to catch her breath. The crowd had thinned out, and she was far away from her neighborhood. She thought she had escaped death. But actually she had been carrying it in her arms all along. She looked down into the glazed, open eyes of her still daughter and fell unconscious.

The Bhopal district collector, Moti Singh, and the superintendent of police, Swaraj Puri, were awakened in the middle of the night by the insistent ringing of their telephones. Singh and Puri were in charge of district administration, the local police department, and civil defense efforts. They rushed at once to the police control room to coordinate emergency relief efforts. But they, along with hundreds of other governmental officials and Union Carbide plant managers, were caught sleeping in more ways than one. Nobody seemed to know what gas had leaked, how toxic it was, or how to deal with the ensuing emergency. The police and the army tried to evacuate affected neighborhoods. They were too slow, and instead of being told to lie on the ground with their faces covered with wet cloths, people were urged to run. 200,000 residents fled in panic into the night.

Morning found death strewn over a stunned city. Bodies and animal carcasses lay on sidewalks, streets, and railway platforms, and in slum huts, bus stands and waiting halls. Thousands of injured victims streamed into the city's hospitals. Doctors and other medical personnel struggled to cope with the chaotic rush, knowing neither the cause of the disaster nor how to treat the victims. Groping for anything that might help, they treated immediate symptoms. They washed the eyes of their patients with water and then soothed the burning with eye drops. They gave the victims aspirin, inhalers, muscle relaxants, and stomach remedies to relieve their symptoms.

Before the week was over, nearly 3,000 people had died. More than 300,000 others had been affected by exposure to the deadly poison. About 2,000 animals had died, and 7,000 more were severely injured. The worst industrial accident in history was over.[1]

But the industrial *crisis* that made the city of Bhopal international news had just begun. Its ramifications were both local and global. As time went on, victims suffering from the long-term health effects of exposure to MIC died. There were continuing controversies over how many people had actually perished and which treatments might be helpful to surviving victims. Family life in Bhopal was radically disrupted, as wives and children with no preparation for life outside the home were forced to go to work and manage the family's financial affairs. The victims sued the government of India and Union Carbide. The government, seeking to protect its own legitimacy, sued Union Carbide in the United States on behalf of the victims. Governments in other countries took steps to stop Union Carbide and other chemical companies from establishing similar plants in their communities. Union Carbide's top executives were arrested upon their arrival in Bhopal, and the corporate prestige and financial health of the thirty-seventh-largest company in the world—a strong and proud corporation with a long history—was dealt a heavy blow.

In one sense, the Bhopal crisis was simply an industrial accident—a failure of technology. But the real story behind the accident goes much deeper than mere technology. It extends to the organizational and socio-political environment in which the accident occurred.

Organizational pressures within Union Carbide contributed to both the accident and the ensuing crisis. The Bhopal plant was an unprofitable operation, for the most part ignored by top Union Carbide officials. With several of Union Carbide's traditionally profitable divisions in the United States faltering, the Bhopal plant was a prime candidate for divestiture. The Indian subsidiary that owned the plant, Union Carbide (India) Ltd. (UCIL), was primarily a battery company that had made an unsuccessful foray into the pesticides market. At the time of the accident, the Bhopal plant operated at only about 30 to 40 percent of capacity and was under constant pressure to cut its costs and reduce its losses.

But it was more than Union Carbide's financial difficulties that set the stage for the crisis. The economic, political, and social environment of Bhopal also played a contributing role. At the time of the accident, Bhopal was a peculiar combination of new technology and ancient tradition sitting in somewhat uncomfortable relation to each other.

Though the city is nearly 1,000 years old, its industrial capacity, until recently, was primitive. In the last thirty years industrial growth was encouraged in Bhopal, but the necessary infrastructure needed to support industry was lacking. There were severe shortcomings in the physical infrastructure, such as supplies of water and energy and housing, transportation and communications facilities, as well as in the social infrastructure, including public health services, civil defense systems, community awareness of technological hazards, and an effective regulatory system.

Nor was industrial growth accompanied by rural development, which might have slowed the migration of people from the hinterlands. The city's population grew at three times the overall rate for the state and the nation in the 1970s. This heavy in-migration, coupled with high land and construction costs, caused a severe housing shortage in the city. For shelter, migrants built makeshift housing, which in turn became slums and shantytowns. By 1984, more than 130,000 people, about 20 percent of the city's population, lived in these slums. Two of these large slum colonies were located across the street from the Union Carbide plant.

Thus, at the time of the accident, several thousand, for the most part illiterate, people were living in shantytowns literally across the street from a pesticide plant. They had no idea how hazardous the materials inside the plant were, or how much pressure the plant was under to cut losses. Indeed, most of them believed it produced "plant medicine" to keep plants healthy and free from insects.

Not all industrial accidents become crises. They trigger crises only when technological problems occur in economic, social, and political environments that cannot cope with them. The Bhopal accident became a crisis not because of technological problems alone but also because of environmental conditions outside the plant. The plant was operated by a company under pressure to make profits and/or cut losses; it was sanctioned by a government under pressure to industrialize, even though the appropriate industrial infrastructure and support systems were missing; and it was located in a city completely unprepared to cope with any major accident. It was these factors, combined with the technological failures that actually caused the accident, that expanded the initial event into a crisis.

Although it was the worst industrial crisis in history, Bhopal-like crises are hardly unusual. All over the world—and, increasingly, in developing countries—industrial crises have become more frequent and devastating. For this reason they deserve close attention. For example, the dioxin poisoning in Seveso, Italy, and the Three Mile Island and Chernobyl nuclear power plant accidents represent industrial crises. These crises present a novel and challenging set of problems for corporations, government agencies, and communities. The causes of crises—in particular, the causes of the secondary effects that turn an accident into a crisis—are difficult to ascertain and remediate because they are so deeply rooted in the various social and economic systems of the countries involved. But failure to meet the challenge will result in more deaths, continued environmental destruction, and a severe downgrading of quality of life.

While it is probably not useful to think about dealing with industrial crises in terms of "solutions"—there is no single solution applicable to all conditions—we can build a greater understanding of who the *stakeholders* are in industrial crises and how their actions can exacerbate or minimize these crises. Bhopal provides a textbook case study for building this kind of understanding.

CHAPTER 2

The Causes and Characteristics
of Industrial Crises

Harish Dhurvey, the stationmaster of the Bhopal railway station, died a hero. On the day of the accident, December 3, Dhurvey had risen while it was still dark to receive a visiting dignitary who was to arrive by special train. The stationmaster soon felt the crippling effects of the poisonous gas. But, exhibiting rare presence of mind, he thought of the danger this situation presented to railway traffic. He knew that trains had to pass by the Union Carbide plant on their way to the Bhopal railway station. Instead of running to save his own life, the stationmaster informed stations before and after Bhopal to stop all rail traffic on the outskirts of the city. Thus, trains did not pass through Bhopal, and the lives of thousands of passengers were saved. But the stationmaster died of MIC inhalation.

In his efforts to save thousands of lives, the stationmaster, whether he knew it or not, was doing his part to prevent a serious industrial accident from escalating into a crisis. As horrendous as the Bhopal accident was, it need not necessarily have become a crisis. Accidents become crises when subsequent events and the actions of people and organizations with a stake in the outcome combine in unpredictable ways to threaten the social structures involved.

Historically, the concept of *crisis* originated in the medical field. It referred to phases of an illness in which the body's self-healing powers became inadequate for recovery, even with external help from life-support systems and medicines. A medical crisis represents the advanced point of a progressively worsening illness, when the illness acquires an objective force against which both patient and doctor are powerless. Such a crisis ends in a structural transformation of the body that may include permanent damage or death. Resolution of the crisis is a liberating experience that restores the physical powers of the patient. Both structural transformation and liberation are integral aspects of crisis resolution.

Crisis in social systems refers to situations that threaten the existing form and structure of the system. If existing social structures are incapable of resolving economic, social, cultural, and political problems, the system's integration is threatened and it faces crisis.[1] Industrial crises are man-made disasters caused by industrial activities. Some dysfunctional effects of industrial activities can cause large-scale physical and social harm, resulting in an industrial crisis. Physical harm includes deaths and injuries to humans and animals and environmental destruction—the pollution of water, soil, and air. Social harm includes economic losses to individuals, corporations, and public-sector agencies; disruption of social relationships and cultural arrangements; and political upheaval.

Industrial crises take many forms, but they are always triggered by a specific event that is identifiable in time and place and traceable to specific man-made causes—in short, a *triggering event*, such as the gas leak from the Union Carbide plant in Bhopal on December 3, 1984. Significantly, not every industrial accident leads to a crisis, and not every industrial crisis is the result of a deadly industrial accident.

When the reactor core at the Three Mile Island Nuclear Power Plant nearly melted down in 1979, no one was killed, but the incident, nonetheless, clearly precipitated a crisis. The accident was triggered by a fault in the secondary cooling system that escalated to the primary cooling system and eventually resulted in a near meltdown. Fifty thousand residents were evacuated, and social disruption lasted for weeks. Metropolitan Edison, the owner of the plant, incurred a cost of nearly $4 billion as a result of the accident.[2] President Carter visited the accident site twice and instituted a special commission of inquiry. Regulatory and insurance industry responses to the accident completely changed the face of the American nuclear power industry.[3]

Consider the contrasting example of transportation accidents, which are responsible for 250,000 deaths worldwide each year. These accidents have never led to a crisis because the public at large does not perceive them as such. Despite the large number of deaths, the social structure surrounding automobiles—the everyday reliance on the car as well as the economic power of automobile makers—faces no serious threat.

Types of Triggering Events

There are different kinds of triggering events, which can occur in both the production and the consumption of technologically based products (see Figure 2–1). Some of the most commonly occurring types are: industrial accidents, environmental pollution, product injuries, and product sabotage.

Industrial Accidents

Many of the most tragic and dramatic industrial crises have been triggered by some kind of accident within an industrial plant. Bhopal was the worst in history, but there are other notable examples. In the crisis at Chernobyl in the Soviet Union, the triggering event was a fire at the nuclear power plant. The worst industrial accident in U.S. history occurred near Galveston, Texas, in 1947, when a freighter loaded with ammonium nitrate exploded after a fire broke out on board. The explosion destroyed a nearby chemical plant, set off smaller explosions and fires in the area, and killed 576 people.

There is no question that the twentieth century has produced a rapid rise in the rate of major industrial accidents and deaths. Half of the century's twenty-eight most serious industrial accidents (those killing at least fifty people) have occurred since 1977 (see Table 2–1). Three of these accidents, including the Bhopal disaster, occurred in 1984 alone – all in rapidly industrializing countries. Industrial accidents are more common in developing nations because they most often lack an adequate industrial infrastructure.

Evacuations caused by industrial accidents follow a similar pattern. Of the 919,000 people evacuated in major rescue efforts (2,000 or more people) between 1967 and 1984, more than 90 percent were evacuated in incidents that have taken place since 1979. More than 50 percent of that figure were evacuated in the year 1984. Figure 2–2 illustrates these rising trends.

On a slower and less dramatic scale, crises occur in the industrial workplace because of occupational health and safety hazards.[4] Workplace injuries and disease are responsible for nearly one million deaths worldwide each year. As new technologies are incorporated into the workplace, workers are exposed to new forms of health risks. Employees have limited choices in escaping hazardous work environments and lack control over working conditions.[5]

Figure 2-1. Types and Examples of Industrial Crises.

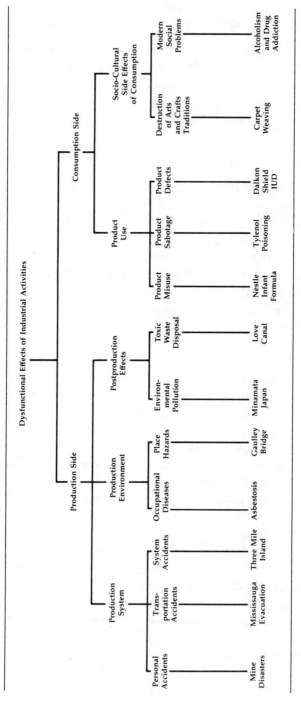

Table 2–1. Major Industrial Accidents (causing more than 50 deaths each).[a]

	Place	Cause	Deaths	Other Damages
1907	Pittsburgh (United States)	Explosion in a steel works	59	Several persons unaccounted for
1917	Petrograd (USSR)	Factory explosion	100	
1921	Oppau (Germany)	Fertilizer factory explosion (ammonium nitrate)	561	1,900 casualties, damage to the town
1933	Neuenkirchen (Germany)	Gas explosion in steel works	63	Several hundred casualties, 70 houses destroyed
1939	Zarnesti (Romania)	Leak of 25 tn of chlorine in a factory	60	300 casualties
1942	Tessenderloo (Belgium)	Explosion in a chemicals plant (ammonium nitrate)	200	1,000 casualties
1943	Ludwigshafen (Germany)	Factory explosion of 16.5 tn of butadiene	57	439 casualties
1944	Cleveland (United States)	Explosion of 4,300 m^3 of confined liquefied natural gas, fire ball	136	350 casualties, streets swept by burning gas, windows broken; 79 houses, 2 factories, and 79 cars destroyed ($6.8 m)
1947	Texas City (United States)	Explosion of a ship with a cargo of ammonium nitrate (1,750 tn)	532	200 unaccounted for, 300 casualties, serious damage to city

Table 2-1 continued.

	Place	Cause	Deaths	Other Damages
1948	Ludwigshafen (Germany)	Explosion of confined dimethyl ether	245	3,800 casualties, damage at 8 km distance (French francs 80 m)
1948	East Germany	Steam-generating station explosion. Pulverising coal	50	76 casualties
1956	Minamata (Japan)	Mercury discharge into river and bay	250	Over 100,000 alleged mercury-poisoning casualties
1957	Barhain	Cotton/wool explosion	57	
1970	Osaka (Japan)	Explosion of confined gas in an underground railway construction site	92	
1978	Los Alfaques (Spain)	Explosion of liquefied propylene in transport by lorry	216	200 casualties (French francs 144 m compensation)
1978	Xilatopec (Mexico)	Explosion of 10,000 liters (l) of liquid natural gas (LNG) following multiple pileup involving lorry and 12 vehicles	100	150 casualties
1978	Huimanquilla (Mexico)	Gas pipeline fracture	58	
1979	Istanbul (Turkey)	Ship/tanker collision	55	95,000 tn of oil on fire
1979	Bantry Bay (Ireland)	Exposion of tanker at berth	50	

Year	Location	Event	Deaths	Comments
1979	China	Offshore-rig collapse	72	
1980	Norway	Offshore-rig collapse	123	
1980	Alaska	Fire at oil-rig construction	51	
1980	Canada	Offshore-rig collapse	84	
1982	Tacoa (Venezuela)	Oil explosion and fire at power station	145	Fire in neighborhood, 1,000 casualties
1984	Cubatao Sao Paulo (Brazil)	Petrol explosion following pipeline fracture	508	Fire in a shanty town (3,000 inhabitants) built illegally on Petrobras lands (Petrobras claimed no more than 90 deaths)
1984	San Juan (Ixhuatepec, Mexico)	Explosion of LNG reservoirs (90,000 barrels)	452	Fire in shanty town (4,248 casualties, 31,000 homeless, 300,000 evacuated, flames 300 m high, 300 houses destroyed)
1984	Bhopal (India)	Emission of 40 tn of methyl isocyanate	2,500	About 180,000 other casualties. 300,000 people left the area voluntarily when the plant was recommissioned.
1984	Gahri Ohoda (Pakistan)	Explosion of a natural gas pipeline	60	

a. Not including accidents in the USSR, accidents involving explosives and munitions, mining accidents and gas distribution accidents, and transportation accidents (passengers).

Sources: Patrick Lagadec, *Major Technological Risk* (London: Pergamon Press, 1981); Henri Smets, "Compensation for Major Exceptional Environmental Damage Caused by Industrial Activities" (Paper delivered at the conference on Transportation, Storage and Disposal of Hazardous Materials, IIASA, Laxenburg, Austria, July 1–5, 1985).

Figure 2–2. Industrial Accident Death and Evacuation Trends.[a]

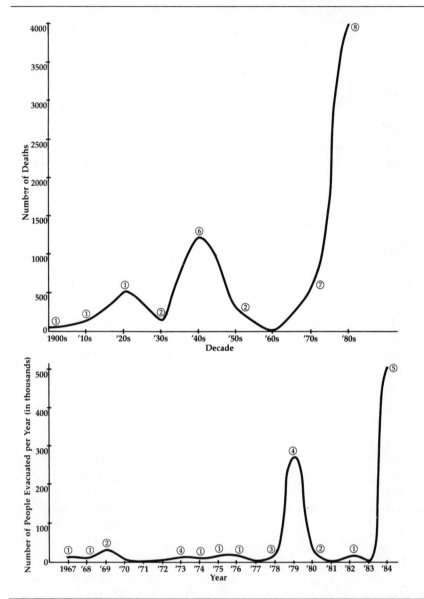

a. Only industrial accidents causing more than fifty deaths are listed. Circled figures represent number of accidents and evacuations.

Source: Adapted from Henri Smets, "Compensation for Exceptional Environmental Damage Caused by Industrial Activities" (Paper delivered at the conference on Transportation, Storage and Disposal of Hazardous Materials, IIASA, Laxenburg, Austria, July 1–5, 1985).

In the long run, these conditions can cause thousands of workers to contract life-threatening disesases and thus lead to industrial crises.[6]

Environmental Pollution

Industrial crises may also be triggered by environmental pollution incidents. Triggering events of this sort may occur suddenly or gradually.[7] In a single year, acid rain and air pollution killed 10 percent of all of Switzerland's spruce trees and 25 percent of all its fir trees. By contrast, the Love Canal crisis in New York State took decades to develop. Toxic wastes were dumped on the site in the 1940s and '50s, but it was not until 1976 that reports of groundwater contamination triggered the crisis.[8] In the developing world, environmental pollution is compounded by widespread poverty. The lack of adequate housing, water supplies, public health services, and pollution-prevention measures has led to an unprecedented environmental crisis.[9] In India, 70 percent of all groundwater is polluted, and waterborne diseases are endemic.[10] Moreover, more polluting industries are located in developing countries.[11]

Cubatao, a town of nearly 100,000 people, located on the southern coast of Brazil, is in a continuous state of crisis because of environmental pollution. It has the world's highest level of pollutants in rain, groundwater, and air. One thousand tons of toxic gases and particles are emitted into the air *every day* in a thirty-square-mile area housing some 111 industrial plants. A government study showed that the rainwater contained high concentrations of sulphur dioxide and sixteen other pollutants. Of these, six pollutants (ammonia, iron, phosphates, flourites, calcium, and sulphates) were at the highest level known in the world. One hundred times during 1984, air pollution in Cubatao reached 240 micrograms (mcg) of chemical dust per cubic meter (pm^3) the threshhold level for long-term health damage. Pollution often exceeded the *alert* level of 475 mcg, and twelve times during the year it exceeded the *emergency* level of 875 mcg. When air pollution is very high, children and elderly people are given emergency supplies of oxygen. The city has some of the highest rates of cancer, tuberculosis, asthma, still births, and birth defects attributable to pollution in the world. Around the city, vegetation has virtually ceased, causing soil erosion and landslides. Rivers in the area cannot support life.[12]

Product Injuries

Product injuries are a relatively recent, but steadily increasing, catalyst of industrial crises. Over the last few years, the use of the following products has triggered crises in the United States: the Dalkon Shield intrauterine devices; the Rely Tampon; food products, such as Jalisco cheese and Riunite wines; and chemicals, such as Agent Orange and asbestos. Each of these products has been associated with multiple injuries, diseases, or deaths to users. Consumers injured by these products have sued for damages and have received increasingly large compensations. As Figure 2–3 indicates, the number of liability lawsuits arising from product injuries is growing.[13]

The misuse of nondefective products occurs when consumers ignore product-use instructions, because the instructions are either inconvenient or impractical. Misuse usually causes minor injuries and rarely reaches crisis conditions. However, if misuse is not checked or is actively promoted by manufacturers, it can assume crisis proportions. Two primary examples of this include pesticide poisoning and, in developing countries, infant formula misuse.

Estimates of death and poisoning by pesticides vary widely. The World Health Organization has estimated that 5,000 people are killed and 500,000 poisoned by pesticides each year.[14] More recent studies have estimated that 10,000 people die each year from causes related to pesticide production alone.[15]

Pesticide misuse also indiscriminately destroys vegetation and animal life and contaminates water sources. In cotton-growing areas of Peru, bird life was entirely eliminated because the areas were excessively sprayed with pesticides.[16] Farmers overuse pesticides in the hope of quickly eliminating large numbers of insects, but overuse leads to the development of more chemical-resistant strains of pests and the contamination of food and fiber products.[17] Users themselves are also adversely affected by overuse. One study revealed that up to 40 percent of the workers in spraying operations suffered from pesticide poisoning.[18]

Controversy over the use of infant formula in developing countries began in the early 1970s, when thousands of infants weaned on formula in those countries died or were found to be malnourished. Since the basic product was considered safe, manufacturers

Figure 2-3. Product Liability Suits and Jury Verdicts.

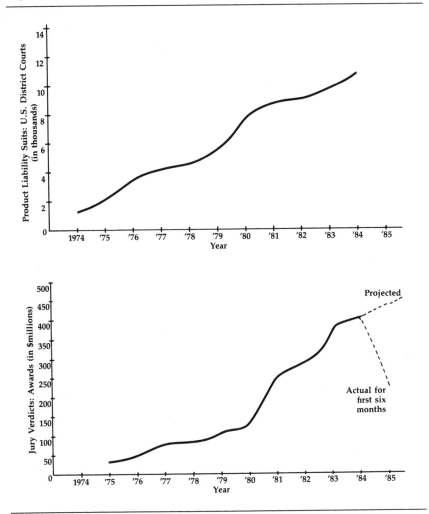

Sources: Administrative Office of U.S. Courts, Washington, D.C.; jury verdict; Research Inc., Solon, Ohio.

would not acknowledge that a problem existed until consumer activists forced them to do so. Studies showed that the infant formula did not provide the immunological properties of breast milk. Moreover, in developing countries, the formula was mixed with contaminated water and diluted below suggested nutritional levels to save money. These practices turned a safe product into a harmful one. Aggressive marketing by manufacturers proliferated the hazard.

The Nestle Company became the prime target for consumer activists because it had 50 percent of world market share of infant formula. In 1977 activists initiated a boycott of Nestle products; the boycott cost the Nestle company millions of dollars in lost sales. The crisis, which lasted well over a decade, involved extended conflicts among the multinational companies that manufactured the product, their host and home country governments, consumer activist groups around the world, church agencies, and international agencies such as WHO and UNICEF.[19]

Product Sabotage

An even more frightening source of industrial crises is the sabotage of consumer products. The best-known example is the lacing of Tylenol with cyanide and strychnine. In 1982, five people in the Chicago area died after consuming cyanide-laced capsules. Subsequently, the Food and Drug Administration initiated nationwide testing of Tylenol and discovered strychnine-laced capsules in California. Before the tragedy ended, Johnson & Johnson, the manufacturer, had recalled 31 million bottles of its products at a cost of more than $100 million. Affected consumers sued the company, and its market share dropped from 35 percent to only 8 percent. The mystery of these poisonings was never resolved.

Tylenol was placed back on the market with triple tamper-resistant packaging and within eight months regained its market share and reestablished itself as the leading over-the-counter painkiller. However, in 1986, another cyanide-laced capsule killed one person in Yonkers, New York. The incident created nationwide public alarm and constituted a major public health risk. In response the company initiated another national recall of millions of capsules, and announced plans to abandon the manufacture of Tylenol in capsule form altogether. Coping with the crisis cost the company over $150 million.

In Japan, soft drinks sold at a vending machine were found to be contaminated with paraquat, a deadly pesticide freely available in Japan. Several people died and many were injured as a result. Police later discovered that the poisoned cans had been planted in the dispensing cage of vending machines. Customers picked up these cans, presuming that the machine had, by mistake, dispensed two cans instead of one.

Product sabotage, though still uncommon, raises frightening questions about our ability to protect ourselves from terrorists and psychopaths. Consumers are extremely vulnerable targets because of the ease with which mass-consumed products can be altered. Product sabotage can also be used to inflict severe financial and market share losses on corporations.

Analyzing Industrial Crises

In many ways, industrial crises are not unlike natural disasters. Sociological studies of natural disasters (floods, earthquakes, hurricanes, storms) have described them as unplanned and socially disruptive events with sudden and severe destructive effects. The three core properties of natural disasters are (1) they can be specifically identified in *time* and in *place*, (2) they affect *social units*, and (3) those social units take *responsive actions* to mitigate the disaster's consequences.[20] Even though the precise time, place, and severity of natural disasters cannot be predicted, their effects can be mitigated both before and after they occur.[21]

Industrial crises can be understood in terms of these same properties. But in industrial crises, the relationship between the technological, organizational, and social causes of the crisis and the existing social order is particularly important. Social agents, institutions, and relationships (rather than nature) are responsible for the triggering event in industrial crises, and social processes are the vehicles by which crises proliferate. Dealing with the consequences invariably involves social and political conflicts. Thus, industrial crises are fundamentally shaped by existing social order and class relations.[22]

Industrial accidents are more complex than natural disasters. A natural disaster is a single event over which no human being has control. But an industrial crisis is a complex system of interdependent events and involves multiple, conflicting stakeholders. The Bhopal accident escalated to a crisis because of the actions of the government of India, Union Carbide, and many other stakeholders, both before and after the accident. Thus, to understand how industrial accidents become crises, it is extremely important to examine closely the triggering events and responsive actions that follow them.

There is a close connection between the level of risks associated with technological systems and the physical and social environments in which these systems function. The quality of the industrial infrastructure—water, energy, public health systems, transportation, communications, educational institutions, the available labor force—influences the probability of accidents occurring in the first place, and then the chances of it escalating into crisis.[23] Technological systems supported by a weak infrastructure, as was the case at the Bhopal plant, run much higher risks of accidents, environmental pollution, occupational health hazards, and industrial sabotage. Often, the ability of stakeholders to respond effectively to the accident—or to a worsening crisis—depends on the strengths or weaknesses of the local infrastructure. Figure 2–4 lays out a framework for examining such industrial crises.

Industrial crises are most likely to occur in areas that are industrializing rapidly, but lack strong supporting infrastructure. Technological systems operating in areas where the infrastructure is weak possess the preconditions for crisis, manifested in chronic problems with accidents, environmental pollution, and occupational diseases. When these problems are exacerbated by human, organizational, and technological failures, they can trigger industrial crises.

These conditions are most likely to exist in developing countries. That is why even though only 9 of the 28 worst industrial accidents of the century (the ones in which 50 or more people died) have occurred in developing countries, those 9 accidents caused more deaths than the remaining 19. The 19 serious accidents in industrialized countries led to the deaths of just under 3,000 people. But the 9 in developing countries caused the death of more than 4,000 people.[24]

The damage caused by recent industrial crises has also weighed heavily on developing countries. More than 500 people were killed when gasoline leaking from a pipeline exploded in a shantytown in Sao Paolo, Brazil, in February of 1984. And just two weeks before the Bhopal accident, about 500 people died when a liquified gas storage plant exploded in Mexico City.

However, industrialized countries are not immune to industrial crises. The Three Mile Island and Chernobyl nuclear disasters and

Figure 2–4. Framework for Industrial Crises.

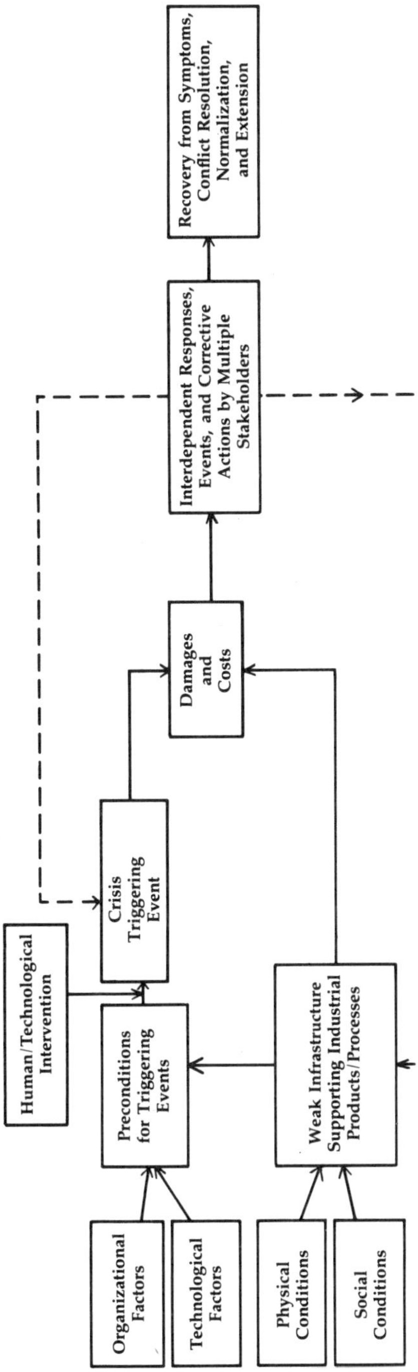

the explosion of the Space Shuttle *Challenger* all occurred in highly industrialized countries and under the direction of technologically sophisticated organizations. Similarly, crises related to industrial pollution, such as the Minamata disease in Japan and the toxic waste disposal crisis in the United States, are also prevalent in industrialized countries. While these crises may not kill many people, they are more materially destructive and more costly than industrial crises in developing countries.[25]

Interdependent Events and Multiple Stakeholders

Crises are constituted of interdependent events that take place in geographically dispersed locations and at different times. Typically, simultaneous chains of events occur in different arenas—for example, the relief arena, the technological arena, and the legal arena. Events within a chain are causally linked, but they are difficult to anticipate and control.

Parallel chains of events affect each other indirectly, but they are only loosely coupled, or connected.[26] Loose coupling occurs because there is no single centralized agency responsible for crisis management. In addition, different individuals and agencies are in charge of each chain of events, which leads to poor coordination and information exchange. For example, legal proceedings in U.S. courts aimed at procuring compensation for the Bhopal victims indirectly affected relief work in Bhopal. The American judge presiding over the case sanctioned $5 million for "interim relief." However, legal proceedings were only loosely connected with the relief work, which meant that there was no follow-up on how this money was used. The money intended for immediate interim relief did not reach victims for more than a year after it was ordered by the judge.

Industrial crises are also characterised by the presence of multiple stakeholders. Corporations such as Union Carbide, which own or manage the industrial plant where the triggering event takes place, are always major stakeholders. These corporations are legally liable for damages caused by their products, accidents, or hazards emanating from their premises. In some countries, such as the United States, legal liability extends to other corporations, such as equipment manufacturers, design and engineering consultants, and raw material suppliers, who are jointly liable for damages caused by their products or designs.

Government agencies in charge of the industrial and social infrastructure – regulations, civil defense, and public health – are stakeholders in two respects. They help mitigate the effects of crises and provide regulatory and monitoring services to prevent similar crises. Failure to perform these tasks adequately threatens the government's own legitimacy and makes the crisis a political liability. In protecting its legitimacy, the government does not always serve the public interest. It acts in self-defense to retain its legitimacy and power.

The nuclear accident at Chernobyl is an excellent example of a government acting self-defensively. When a fire erupted at the nuclear plant in April of 1986, the Soviet government made no public announcement of the event. But the radiation leaking from the plant quickly spread across international borders. It was not until increased levels of radiation were detected in Sweden, some 700 miles away, that citizens in potentially affected areas even knew the fire had occurred. Even after government agencies acknowledged that the accident had taken place, they still refused to reveal much information about the aftereffects, leading media around the world to report exaggerated estimates of damages. Ultimately, the strategy of revealing little information – a strategy designed to protect the government's legitimacy – called that very legitimacy into question.

Another set of stakeholders is the public and public interest groups. They provide relief services and exert pressure on corporate and government organizations to aid victims in recovering from damages. Unresolved crisis problems and the persistent failure of existing organizations to cope with them, erodes mass loyalty to the state, and makes people lose faith in establishment organizations. Autonomous public groups emerge to mitigate the crisis and put pressure on state and international agencies to resolve problems. The rise of social movements such as environmentalism and consumerism in the 1970s can be seen as a manifestation of eroding mass loyalty, particularly since the existence of such organizations has been fueled largely by crises surrounding environmental degradation and product harm.

The most profoundly affected stakeholders – and ironically, sometimes the most easily forgotten because of their powerlessness – are the victims. These include workers in production facilities, consumers, and residents of communities where hazardous facilities are located. Sometimes even remote observers of crisis events suffer significant effects. Because of genetic or delayed medical effects even unborn children of persons affected by crises are potential victims.

While not one of the primary stakeholders, the media plays an important role in communicating crisis events to the public. It shapes public perceptions and responses to crises. But despite its enormous resources, the media's coverage of crisis events is most often fragmented and equivocal and frequently lacks objective data. This is a result of both, stakeholders' attempts to control communication and a genuine lack of information about the cause of the crisis. Such partial and distorted coverage gives rise to myths, false alarms, and heightened perceptions of harm.[27]

The Inherent Contradictions in Coping with Industrial Crises

When a triggering event occurs, spontaneous reactions by different stakeholders solve some of the immediate problems, but they also create new problems—thus prolonging the crisis and making it worse. The crisis expands to affect people and organizations in remote locations. It spawns new conflicts. Crisis responses by organizations and citizens deal, for the most part, with the symptoms of a crisis without eliminating its fundamental causes. Hence, crisis potential continues to exist.

This is exactly what happened in Bhopal, and it is, to a certain extent, an inevitable result of the economic, social, and political environment in both industrialized and developing countries. In both situations, the environment fosters the conditions in which crises occur, and it determines the capacity of communities to mitigate the effects of crisis.

Many industrialized countries, such as the United States and its allies, have capitalist systems of economic production and democratic welfare state systems of political governance. Other countries have some form of state capitalism and socialist governance structures. In both cases, three interdependent and conflicting subsystems constitute the environment of industrial crises:

1. The *economic subsystem*, consisting of private or public sector corporations
2. The *political-administrative subsystem of the state*—that is, the government
3. The *system of normative societal structures* that govern socialization processes

The state receives fiscal support from the economic system and provides it with regulatory services. It also provides services through a heterogenous set of political and administrative institutions that manage the economy and the normative structures of society.[28]

But embedded in this organization of society is a fundamental contradiction that obstructs the state's ability to deal with industrial crises. On the one hand, the state must create conditions for capital to be invested in the most productive and profitable manner. On the other hand, it must regulate this investment without alienating owners, in an effort to preserve public goods, such as the environment and to ensure an equitable distribution of wealth.[29]

Thus, the state must simultaneously control and free the productive enterprise system that is the source of industrial crises. This contradiction prevents the state from being an effective regulator of crises. Even in rich countries that ostensibly have the resources to deal with potential crises, enough resources are not allocated to prevent crises.

The contradiction in developing countries is even more pronounced. Typically, the state controls a major part of industrial production. Thus, the state imposes contradictory pressures on itself, to create a highly productive industrial system and simultaneously, to limit that system's destructive effects. In many developing countries, such as Brazil, India, and Mexico, public-sector manufacturing plants account for a large proportion of industrial production and employment. They also cause the most environmental pollution and account for the largest number of deaths caused by industrial accidents. (In the Mexico City tragedy of 1984, the liquified petroleum gas storage facility was operated by the government-owned PEMEX oil company.) The underlying contradiction in the state's role in industrial crises goes unnoticed, because the state—through its record-keeping procedures, influence over the media, and direct supervision of the crisis—controls the dissemination of information concerning the crises.[30]

Even when the government does not directly own production plants, often it plays a major role in encouraging and establishing them through its industrial policies.[31] In Bhopal, the state and central governments encouraged companies such as Union Carbide to set up industrial operations and, in fact, overruled the city government's objections to locating a plant handling hazardous materials on that particular site.

Rapid Industrialization

But the causes of industrial crises are not entirely rooted in political and socio-economic systems. The rapid pace of industrialization — eagerly sought by many developing nations, and encouraged and supported by the industrialized ones — has also contributed to the problem.

Many developing countries gained independence from their colonial rulers only after World War II. The colonial parents regarded them as providers of raw materials and markets for manufactured goods, so they made little effort to build up their industrial production capacity or infrastructure.

After gaining independence, many developing countries sought to build their industrial capacity quickly by acquiring technology from industrialized nations. These efforts were encouraged by the developing countries' former colonial rulers, facilitated by the existence of a Western-educated ruling elite, and implemented through the policies of international development lending organizations such as the World Bank. As Table 2–2 indicates, between 1960 and 1982 most developing countries showed a significant increase in the percentage of their gross domestic product (GDP) derived from industrial activities. Some countries, such as Indonesia, Nigeria, and Bangladesh, more than doubled this figure. More importantly, in many countries, the industrial percentage of GDP rose high enough to equal or surpass that of the industrialized countries, whose economies were becoming more service oriented.[32]

The two most common vehicles for the transfer of technology were intergovernmental contracts and multinational corporations. Obviously, industrial technology accelerated economic development efforts in these countries by bringing new investments and jobs to countries that had a surplus of labor. But the new technologies really assisted only the small, "modern" sector of developing economies.[33] This sector typically employs only the few people living in urban areas, while the masses remain in rural areas, working in the nonindustrial economy.[34] These industrial technologies also sowed the seeds of future industrial crises.

In their zeal to industrialize, developing countries and their Western partners used foreign investment to build up productive industrial capacity. They did not make commensurate domestic investment in building the infrastructure necessary to support these

Table 2–2. Industrial Output of Selected Countries.

Country	Population mid-1982 (Millions)	Total GDP in $ Millions		Percentage of GDP from Industry	
		1960	1982[a]	1960	1982
Low-income economies					
China	1,008.2	42,770	260,400	33	41
India	717.0	29,550	150,760	20	26
Bangladesh	92.9	3,170	10,940	7	14
Pakistan	87.1	3,500	24,660	16	25
Burma	34.9	1,280	5,900	12	13
Ethiopia	32.9	900	4,010[b]	12	16[b]
Zaire	30.7	130	5,380	27	24
Tanzania	19.8	550	4,530	11	15
Kenya	18.1	730	5,340	18	22
Sri Lanka	15.2	1,500	4,400	20	27
Lower-middle-income economies					
Indonesia	152.0	8,670	90,160	14	39
Nigeria	90.6	3,150	71,720	11	39
Philippines	50.7	6,960	39,850	28	36
Thailand	48.5	2,550	36,790	19	28
Turkey	46.5	8,810	49,980	21	31
Egypt	44.3	3,880	26,400	24	34
Morocco	20.3	2,040	14,700	26	31
Sudan	20.2	1,160	9,290	–	14
Peru	17.4	2,410	21,620	33	39
Cameroon	9.3	550	7,370	–	31
Upper-middle-income economies					
Brazil	126.8	14,540	248,470	35	–
Mexico	73.1	12,040	171,270	29	38
Korea, Rep. of	39.3	3,810	68,420	20	39
South Africa	30.4	6,980	74,330	40	–
Argentina	28.4	12,170	64,450[b]	38	–[b]
Yugoslavia	22.6	9,860	68,000	45	45
Algeria	19.9	2,740	44,930	35	35
Venezuela	16.7	7,570	69,490	22	42
Nicaragua	14.5	2,290	25,870	18	30
Chile	11.5	3,910	24,140	35	34

Table 2–2 continued.

Country	Population mid-1982 (Millions)	Total GDP in $ Millions		Percentage of GDP from Industry	
		1960	1982[a]	1960	1982
High-income oil exporters					
Saudi Arabia	10.0	–	153,590[b]	–	77[b]
Libya	3.2	310	28,360	–	68
Kuwait	1.6	–	20,060	–	61
Oman	1.1	50	7,110	8	–
United Arab Emirate	1.1	–	29,870	–	–
Industrial market economies					
United States	231.5	505,300	3,009,600	38	33
Japan	118.4	44,000	1,061,920	45	42
Germany, Fed. Rep.	61.6	72,100	662,990	58	46
Italy	56.3	37,190	344,580	41	41
United Kingdom	55.8	71,440	473,220	43	33
France	54.4	60,060	537,260	39	34
Spain	37.9	11,430	181,250	–	34
Canada	24.6	39,930	289,570	34	29
Australia	15.2	16,370	164,210	40	35
Netherlands	14.3	11,580	136,520	46	33
East European nonmarket economies					
Romania	22.5	–	55,020	–	57
Hungary	10.7	–	20,710	39	45

a. Figures for 1960 and 1980 based on 1961 net material product.
b. 1981 figures for Argentina, Ethiopia, and Saudi Arabia.

new industries. There are two reasons why infrastructure development does not keep pace with the addition of industrial capacity in most countries: (1) chronic fiscal problems of the state and (2) policy and planning failures.

Fiscal Problems

Financial resources are required to manage the expensive social problems associated with industrialization. But developing countries are not wealthy, and basic services in the areas of health, education, transportation, and energy have taken budgetary preference over environmental, consumer and worker protection, because such services have greater political importance. The fiscal problem is only compounded by the bureaucratic inertia of government agencies in using the limited resources allocated by the state. The gap between problems and the resources available for their resolution grows cumulatively with time, thus perpetuating the potential for industrial crises.

But the problem of adequate resources is not limited to developing countries. Even the richest countries do not allocate sufficient resources to remedy industrial problems. In the United States, for example, the amount of money designated to clean up toxic waste sites is clearly inadequate. Solving the problem could cost up to $100 billion and take up to fifty years. But for the first five years of the government's so-called Superfund program, resources devoted to the effort were $1.6 billion from the federal government and $500 million from private sources. Cleanup work was started on less than 5 percent of the 825 priority sites and completed on only six sites.[35]

Data on plant inspections monitoring health and safety standards reflect the same story.[36] One consequence of meager inspections is a rise in job-related injuries and illnesses among industrial workers. In the United States, these injuries and illnesses per 100 full-time workers in the private sector rose from 7.6 percent in 1983 to 8 percent in 1984. The number of work days lost also rose significantly.[37]

In developing countries, the paucity of resources for coping with the harmful effects of industrial activities is so acute as to be frightening. For example, in India, one of the largest and most populous countries in the world, the central government's entire Department of Environment operated within an annual budget of $650,000 in 1983.[38] In several highly industrialized states of India, each factory inspector is expected to check more than one factory per working day—a requirement that allows barely enough time for even a cursory inspection. In practice, in some states less than 15 percent of monthly inspection targets are met.[39]

Conditions in many other industrializing countries are similar. Ireland, Spain, Mexico, and Romania were found to lack necessary experts in pollution control and environmental protection. The problem is compounded by the fact that the worst environmental offenders are old plants, built before modern pollution-control technology was developed. Requiring these plants to install new pollution-control devices would simply force them out of business and reduce jobs.[40] Governments are reluctant to enforce safety regulations that reduce jobs because this is not a politically wise move.

Table 2-3 provides some indication of the meager resources devoted to environmental protection. Most of the environmental agencies in developing countries are less than a decade old, and their emphasis is on preservation of forests and water resources rather than protection from industrial hazards. Their scientific staff is limited, ranging from 0 to 413 people. Their annual budgets range from $8 million down to $130,000. This lack of commitment to the problems posed by industrial hazards leads to chronic pollution and accidents, such as the one in Bhopal, that escalate into industrial crises.

Policy and Planning Failures

National policy and planning efforts, which are stronger in developing countries than in the industrialized nations of the West, are a powerful means for preventing, containing, and monitoring industrial crises. But they have not been used effectively.

On one level this is attributable to a chronic inability to implement national policy. Disjointed and incremental planning, political pressures, organized resistance to state power, and rapidly changing environmental conditions lead to planning failures.

On another level, national planning and industrial policy, by encouraging the establishment of hazardous industries, can actually contribute to the likelihood of industrial crises. Hazardous technologies have been introduced into communities that have neither the infrastructure nor the resources to handle them safely.

Perhaps the most visible policy failure contributing to industrial crises involves urbanization. All over the developing world, industrial plants are located in populated urban centers. This is

Table 2-3. Environmental Management Resources for Selected Countries.

Country	Nodal Department Established	Staff Total	Staff Scientific	Budget in U.S. $ Millions	Environmental Problems
Low- and middle-income					
Bangladesh	1977	52	120	0.13	Deforestation, urban health
Brazil		228	58	—	Natural resource depletion, industrial pollution
China	1984	1,358	413	—	Deforestation, chemical pollution
Egypt	1982	10	—	8.0	Pollution of Nile by sewage
Ghana	1973	40	10	—	Deforestation, water pollution
India	1980	150	25	0.65	Deforestation, water pollution
Israel	1973	36	30	3.5	Water shortage and pollution
Jordan	1980	17	10	0.15	Acute water pollution
Kenya	1979	100	30	—	Deforestation, soil erosion
Pakistan	1972	113	—	0.30	Unknown
Philippines	1977	100	50	1.00	Deforestation, industrial pollution
Sudan	1975	4	2	0.40	
Venezuela	1980	—	—	—	50% population lives in Caracas slums

Table 2-3 continued.

Country	Nodal Department Established	Staff Total	Staff Scientific	Budget in U.S. $ Millions	Environmental Problems
High income					
Japan	1972	907	374	2,000.00	Industrial pollution, overcrowding
New Zealand	1972	38	16	1.70	Soil erosion, resource depletion
Norway	1972	195	–	113.00	Acid rain, industrial pollution
South Africa	1973	280	216	24.00	Mining pollution
Spain	1972	120	20	10.30	Industrial pollution
Sweden	1967	640	140	55.00	Mercury pollution, acid rain
U.K.	1970	–	–	–	Smoke and river pollution
U.S.A.	1972	12,707	–	4,000.00	Toxic waste disposal, air pollution

Source: *The World Environment Handbook* (New York: World Environment Center, 1984).

largely because most countries' economic development strategies have an urban bias, even though the population is concentrated in rural areas. The investment in industrialization, without a comparable investment in rural development, encourages large-scale migration from rural to urban areas.[41] At the same time, investment in urban housing, transportation, communications, and public health services does not keep pace with the level of investment accorded to industrial plants. Consequently, in highly congested cities like Bangkok, Bombay, Calcutta, Cairo, Lagos, Mexico City, and Rio de Janeiro, industrial plants are located in the middle of established, residential neighborhoods—and, when they are not, they are surrounded by slums and shantytowns populated by rural immigrants who cannot find housing elsewhere.[42]

All the elements of industrial crisis discussed in this chapter were present in Bhopal. Let us now turn to the Union Carbide accident and its aftermath in order to see how these elements helped cause the accident and bring about the disastrous consequences that followed.

CHAPTER 3

Causes of the Bhopal Disaster

In 1984, Union Carbide was one of the most important industrial companies in the world. It was the seventh-largest chemical company in the United States, with both assets and annual sales approaching $10 billion. It owned or operated businesses in forty countries, employing almost 100,000 people. It produced a wide variety of chemical products, including petrochemicals, industrial gases, metals and carbon products, consumer products, specialty products, and technology services.

Despite its size and importance, however, Union Carbide was not considered a blue-chip stock. It faced increasing competition in the chemicals industry, and its financial performance over the previous decade had been lackluster. In 1984, the company's after-tax profit amounted to only about 60 percent of the after-tax profits generated by its chief rivals, Dow Chemical, Du Pont, and Monsanto.[1]

These problems prompted the company to redirect its strategy. It curtailed investments in products that were not doing well— petrochemicals, and metals and carbon products. The petrochemical cutback, in the late 1970s, involved the divestiture of almost three dozen business units and product ventures,[2] and it allowed the company to redeploy funds into divisions that dealt with industrial gases, consumer products, technology services, and specialty products.[3]

Even after this shift in direction, the company suffered major setbacks. The metals and mining division, which supplied products to the declining American steel industry, saw its pretax profits fall from $291 million in 1981 to $31 million in 1984, with little chance predicted for recovery. In 1981, the company introduced a cheaper process for the manufacture of ethylene derivatives that were used as raw materials for polyester, antifreeze, and polyethylene plastics. Instead of driving the competition out of the market, this strategy backfired by creating a worldwide oversupply of ethylene derivatives and driving prices down. The division manufacturing the product dropped from a $131 million profit in 1981 to a $39 million loss in 1982. Even in the normally lucrative area of real estate, Union Carbide had bad luck. In 1977 it lost money by selling its corporate headquarters at one of the few times in history when the Manhattan real estate market was depressed.

These setbacks led the company to plan a major financial restructuring. Union Carbide's cautious management, however, dragged its feet and at the time of the Bhopal accident had still not implemented the restructuring.

Despite its large size and assets, Union Carbide (India) Ltd. (UCIL), like its parent company, faced considerable difficulties. In 1984, UCIL was the twenty-first-largest company in India, with revenues of over Rs 2 billion, or about $170 million. Just over half of UCIL's shares were owned by Union Carbide Corp., but the company also had 24,000 individual shareholders. UCIL operated as part of Union Carbide Eastern, the company's Hong Kong-based international division.[4]

UCIL employed more than 10,000 people, of which almost 1,000 earned Rs 3,000 ($250) per month, making the company one of the best-paying employers in India. It had five operating divisions that served as separate profit centers. Products and facilities of each division are listed in Table 3-1.

UCIL was primarily a company that manufactured and sold electric batteries. In 1984, more than 50 percent of the company's revenues came from this product. But over the years, as its product lines in batteries, chemicals, and plastics matured, the company sought out new markets to maintain its growth. Table 3-2 highlights some key events in the history of UCIL.

The industries UCIL entered were typically technology- and capital-intensive. They required large-scale production and technically skilled labor and catered to mass markets.[5] Most often, UCIL would enter industries still in their early stages of development and gain a dominant position by using the superior technology of its parent company.

One such industry was pesticides. In the 1960s, large-scale use of agricultural pesticides was promoted by the Indian government

Table 3-1. Operating Divisions, Products, and Facilities of UCIL.

Division/Company	Products and Services	1983 Approximate Sales (Rs Million)	Manufacturing Units	Sales Offices
1. Agricultural Products Division	Agricultural products including fungicides, miticides, herbicides, nematocides, Sevidol insecticides, Temik insecticides, Sirmate insecticides	173	Bhopal, R&D facility in Bhopal	Ahmedabad, Bangalore, Calcutta, New Delhi, Guntur, Indore, Pune, Secunderabad
2. Battery Product Division	Eveready and Natex brand batteries, Eveready Commander Jeevan Sathi torches and lanterns, Eveready bulbs and mantles	1,235	Calcutta, Hyderabad, Lucknow, Madras, Srinagar, R&D facility in Calcutta	Ahmedabad, Bangalore, Bombay, Cochin, New Delhi, Gauhati, Jabalpur, Jaipur, Lucknow, Madras, Patna, Secunderabad.
3. Carbon, Metals, and Gases Division	National cinema carbons, Fetiarc gouging electrodes, Emmo photoengraver plates, zinc addressograph strips, electrolyic manganese dioxide, stellite hardfacing rods, electrodes castings and components, process carbons and Fetlarc welding and cutting equipment, carbon electrodes and calot and zinc strips for batteries	80	Calcutta, Madras, Thane	Bombay, Calcutta, Delhi, Madras, Secundrabad

Table 3-1 continued.

Division/Company	Products and Services	1983 Approximate Sales (Rs Million)	Manufacturing Units	Sales Offices
4. Chemicals and Plastic Divisions	Union Carbide polyethylene resin, polythethylene films, acetic acid, butyl alcohol, butyl acetate, ethyl and 2 ethyl hexanol	545	Bombay	Baroda, Bombay, Calcutta, New Delhi, Madras
5. Marine Products Division	Commercial fishing, processing and export of marine products; Visaklapatnam	42	—	—
6. Nepal Battery Company Ltd.	Eveready batteries	—	Balaju	Kathmandu
	Purchased products	25		
Total = 5 Divisions and 1 Company		2,100	13 manufacturing facilities and 2 R&D laboratories	20 Sales offices

Table 3-2. Key Events in UCIL History.

1934	Incorporated as a private company named Ever Ready Company (I) Ltd. on June 20, 1934, with registered office in Calcutta, India. Assembles dry cell batteries.
1940	Production of dry cell batteries started in Calcutta.
1942	Another dry cell battery plant set up in Madras.
1955	Converted into a public company. Starts manufacturing Carbon electrodes.
1958	Flashlight manufacturing plant set up in Lucknow.
1959	Name changed to Union Carbide (I) Ltd.
Early 1960s	Diversifies into chemicals and plastics.
1965	Arc carbon plant set up in Calcutta.
1966	Agricultural Products Division started.
1968	Agricultural Products Division shifted to Bhopal.
1969	Pesticide formulation operations started.
1974	Granted industrial licence to manufacture MIC-based pesticides up to 5,000 tons per annum.
1976	R&D center established in Bhopal for developing new products and doing contract research for Union Carbide Corporation.
1977	Manufacture of MIC-based pesticide commences in Bhopal.
1980	MIC production plant set up in Bhopal.
1982	Joint Venture agreement signed on February 25, 1982, to form Nepal Battery Company Ltd., a project worth Rs 22.5 million.
1984	On December 3, forty-five tons of methyl isocyanate gas leak from a storage tank, killing nearly 3,000 people and injuring 200,000. Bhopal plant closed.

as part of its "green revolution" campaign to modernize agriculture. Pesticides quickly became popular among farmers, and their use tripled between 1956 and 1970.[6]

It was during the 1960s that UCIL entered the market. In the mid-1960s UCIL laid plans for its new Agricultural Products Division, and in 1969 the division began operating the Bhopal plant. The plant was located on the north side of the city, about two miles from the Bhopal railway station and bus stand—the hub of local commercial and transportation activities (see Figure 3-1). Since the plant was initially used only for "formulation" (the mixing of

Figure 3-1. Location of the UCIL Plant.

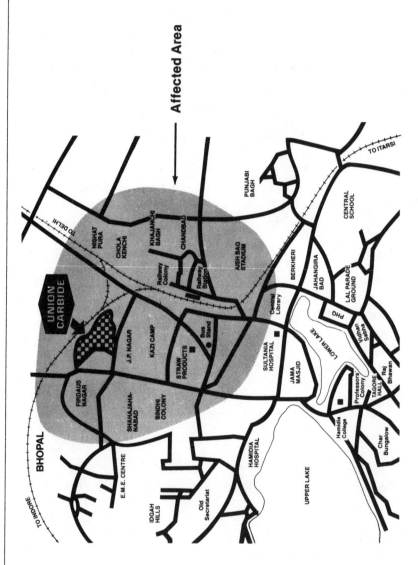

different stable substances to create pesticides), it did not pose a grave danger to surrounding areas.

In 1974, UCIL was granted an industrial license by the central government to *manufacture*, rather than simply formulate, pesticides. By 1977, UCIL had begun producing more sophisticated and dangerous pesticides in which carbaryl was the active agent. Comonent chemicals such as methyl isocyanate (MIC) were imported from the parent company in relatively small quantities.

Within a short period of time, however, the pesticides market became crowded.[7] Fifty different formulations and more than 200 manufacturers came into existence to serve small, regional market niches. This increased competition. Manufacturers were forced to cut costs, improve productivity, take advantage of economies of scale, and resort to "backward integration," that is, not only formulate the final products but manufacture the raw materials and intermediate products as well.[8]

While competitive pressures were mounting, widespread use of pesticides declined.[9] Agricultural production peaked in 1979, declined severely in 1980, and then recovered mildly over the next three years. Weather conditions during 1982 and 1983 were particularly unfavorable, causing farmers to abandon temporarily the use of pesticides.[10] As a result, the pesticides industry became even more competitive in the early 1980s. The expansion and underutilization of production capacity, coupled with a decline in agricultural production, fueled competition.

During this period of industry decline, UCIL decided to "backward integrate" into the domestic manufacture of MIC. In 1979, the company expanded its Bhopal factory to include facilities that manufactured five pesticide components, including MIC. Using this strategy, UCIL hoped to exploit economies of scale and save transportation costs. Establishment of the new, complex technologies required for the domestic manufacture of MIC, however, made the plant much more hazardous than it had been before.

Municipal authorities in Bhopal objected to the continued use of the UCIL plant at its original location. The city's development plan had earlier designated the plant site for commercial or light industrial use, but not for hazardous industries.[11] However, UCIL was a very powerful company in India and in Madhya Pradesh. Its high-paying jobs made it an attractive employer, and among its workers were many former government officials, as well as relatives of high-ranking current officials. The central and state government authorities overruled the city's objections and granted approval of the backward integration plan.

The Accident

At the core of any industrial crisis is a triggering event. In Bhopal, the triggering event was the leakage of a toxic gas, MIC, from storage tanks. Human, organizational, and technological failures in the plant paved the way for the crisis that ensued.

MIC is a highly toxic substance used for making carbaryl, the active agent in the pesticide Sevin.[12] It is highly unstable and needs to be kept at low temperatures. UCIL manufactured MIC in batches and stored it in three large underground tanks until it was needed for processing. Two of the tanks were used for MIC that had met specifications, while the third stored MIC that had not met specifications and needed reprocessing.[13]

A schematic layout of the storage tanks and the various pipes and valves involved in the accident is shown in Figures 3–2 and 3–3. MIC was brought into the storage tanks from the MIC refining still through a stainless-steel pipe that branched off into each tank (see Figure 3–2). It was transferred out of the storage tanks by pressurizing the tank with high-purity nitrogen. Once out of the tank, MIC passed through a safety valve to a relief-valve vent header, or pipe, common to all three tanks, which led to the production reactor unit. Another common line took rejected MIC back to the storage tanks for reprocessing and contaminated MIC to the vent-gas scrubber for neutralizing. Excess nitrogen could be forced out of the tank through a process pipe that was regulated by a blow-down valve. Though they served different purposes, the relief-valve pipe and the process pipe were connected by another pipe called the jumper system. This jumper system had been installed about a year before the accident to simplify maintenance.

The normal storage pressure, maintained with the aid of high-purity nitrogen, was 1.0 kilogram per square centimeter (kg/sq cm). Each storage tank (see Figure 3–3) was equipped with separate gauges to indicate temperature and pressure, one local and the other inside a remote control-room. Each tank also had a high-temperature alarm, a level indicator, and high- and low-level alarms, as shown in Figure 3–3.

Figure 3-2. Schematic Layout of Common Headers of MIC Storage Tanks.

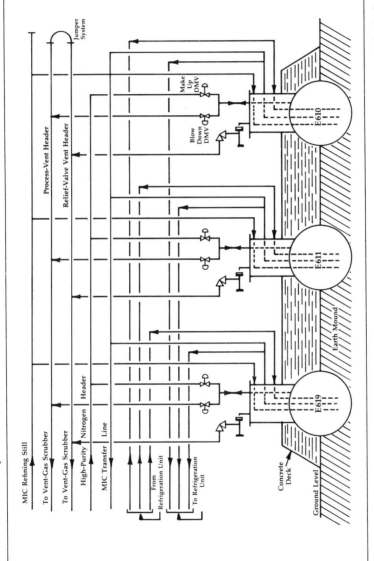

Source: Union Carbide (India) Ltd., *Operating Manual Part II: Methyl Isocyanate Unit* (Bhopal: Union Carbide (India) Ltd., February 1979).

Figure 3–3. MIC Storage Tank.

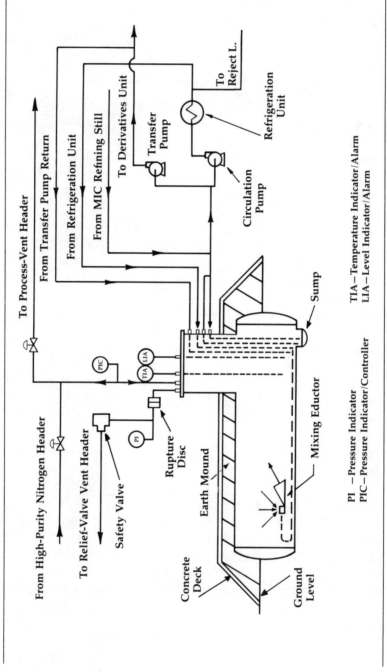

Source: *Bhopal Methyl Isocyanate Incident Investigation Team Report* (Danbury, Conn.: Union Carbide Corporation, March 1985).

The safety valve through which MIC passed on its way to the Sevin plant operated in conjunction with a mediating graphite rupture disk, which functioned like a pressure cooker—holding the gas in until it reached a certain pressure, then letting it out. The rupture disk could not be monitored from a remote location. Checking it required frequent manual inspection of a pressure indicator located between the disk and the safety valve.[14]

The plant had several safety features. The vent-gas scrubber was a safety device designed to neutralize toxic exhausts from the MIC plant and storage system. Gases leaving the tank were routed to this scrubber, where they were scrubbed with a caustic soda solution and released into the atmosphere at a height of 100 feet or routed to a flare.

The gases could also be routed directly to the flare without going through the scrubber. The flare tower was used for burning normally vented gases from the MIC section and other units in the plant. Burning would detoxify the gases before venting them into the atmosphere. However, the flare was not designed to handle large quantities of MIC vapors. A few weeks before the accident the scrubber was turned off to a standby position.

Two additional features of the plant were relevant for safety. The first was a refrigeration system, used to keep MIC at low temperatures, particularly in the summer when the ambient air could reach temperatures as high as 120°. The refrigeration system was a thirty-ton unit that used Freon to chill salt water, the coolant for the MIC storage tank. However, the refrigeration system was shut down in June of 1984, and its coolant was drained for use in another part of the plant, thus making it impossible to switch on the refrigeration system during an emergency.[15] The second important feature was a set of water-spray pipes that could be used to control escaping gases, over-heated equipment, or fires.

The last batch of MIC manufactured before the accident was produced between October 7 and October 22, 1984. At the end of the manufacturing cycle, one storage tank, called tank E610, contained about 42 tons of MIC, while the second tank, E611, contained about 20 tons. After the MIC production unit was shut down, parts of the plant were dismantled for maintenance. The flare tower was shut down so that a piece of corroded pipe could be replaced.

On October 21, nitrogen pressure in tank E610 dropped from 1.25 kg/sq cm, which was about normal, to only 0.25 kg/sq cm.

Because the first storage tank lacked sufficient pressure, any MIC needed in the manufacturing process was drawn from the other tank, E611. But on November 30, tank E611 also failed to pressurize because of a defective valve. Plant operators attempted to pressurize tank E610 but failed, so they temporarily abandoned it and, instead, repaired the pressure system in tank E611.

Then, on the night of December 2 and the early morning of December 3, a series of human and technological errors caused the water used to flush pipes to gush through several open valves and into the MIC tank. There, the water combined with the MIC to produce a hot and highly pressurized mixture of gas, foam, and liquid, which escaped through the plant's stack into the atmosphere.

In the normal course of operation, water and MIC react with each other in small quantities in the plant's pipes, creating a plastic substance called trimer. Periodically, the pipes were washed with water to flush out all the trimer that had built up on pipe walls. Because the mixture of water and MIC was so volatile, the pipes were blocked off with a physical barrier, known as a slip blind, to prevent the water from going into the storage tank.

On the evening of December 2, the second-shift production superintendent ordered the MIC plant supervisor to flush out several pipes that led from the phosgene system through the MIC storage tanks to the scrubber. Although MIC unit operators were in charge of the flushing operation, insertion of the slip blind was the responsibility of the maintenance department. According to workers, the position of second-shift maintenance supervisor had been eliminated several days earlier, and no worker had yet been given responsibility for inserting the slip blind. The flushing operation began at 9:30 P.M.

Because several bleeder lines, or overflow devices, downstream from the flushing were clogged, water began to accumulate in the pipes. Many of the valves in the plant were leaking, including one that was used to isolate the lines being flushed, so water rose past that valve and into the relief-valve pipe. When the operator noticed that no water was coming out of the bleeder lines, he shut off the flow, but the MIC plant supervisor ordered him to resume the process.[16]

Because the relief-valve pipe was about twenty feet off the ground, the water flowed downhill toward tank E610. First it flowed through the jumper system to the process pipe. From that pipe, which is normally open, the water flowed to the blow-down valve, which should have been closed. However, the blow-down valve is part of the system used to pressurize the tank with nitrogen—the same tank whose pressurization system had not been working for weeks. It is possible that this valve had been inadvertantly left open or was not sealed properly.

With the blow-down valve open, about 1,100 pounds of water flowed through another isolation valve, normally left open, and entered tank E610, where it began to react with the MIC being stored there. At 10:45 P.M., a change of shift took place. At 11 P.M., Suman Dey, the new control-room operator, noticed that the pressure in tank E610 was 10 pounds per square inch (psi), well within the operating range of 2–25 psi.

One-half hour later, however, a field operator noticed a leak of MIC near the scrubber. Workers inspected the MIC structure and found MIC and dirty water coming out of a branch of the relief-valve pipe, on the downstream side of the safety valve. They also found that another safety valve, called the process-safety valve, had been removed, and the open end of the relief-valve pipe had not been sealed for flushing. They informed the control room about this. By 12:15 A.M., Dey saw that the pressure in tank E610 had risen to between 25 and 30 psi and was still rising. Within 15 minutes it showed a reading beyond 55 psi, which was the top of the scale.

Dey ran to the tank. He heard a hissing sound from the safety valve downstream, indicating that it had popped. Local temperature and pressure gauges showed values beyond their maximums of 25° C (77° F) and 55 psi. Dey heard loud rumbling and screeching noises from the tank and felt heat radiating from it. He went back to the control room and tried to switch on the scrubber, which had been in a standby mode since the last MIC manufacturing run. But his instruments indicated that the caustic soda was not circulating within the scrubber. In the meantime, field operators saw a cloud of gas gushing out of the stack.

Supervisors notified the plant superintendent, who arrived immediately, suspended operation of the MIC plant, and turned on the toxic-gas alarm to warn the community around the plant. A few minutes later the alarm was turned off, leaving only the in-plant siren to warn workers inside the plant. Operators turned on the firewater sprayers to douse the stack, the tank mound, and the relief-valve pipe to the scrubber. The water spray did not reach thè gases, which were being emitted at a height of 30 meters. The supervisors tried to turn on the refrigeration system to cool the tanks, but since the coolant from the system was drained, the refrigerator could not work. The safety valve remained open for two hours. A mixture of gases, foam, and liquid escaped at a temperature in excess of 200° C (close to 400° F) and a pressure of 180 psi.[17]

Because the plant was so close to the slums, many thousands of people were affected by exposure to this lethal mixture. Nearly 3,000 people died. Thousands more were harmed in some way, many of whom experienced illnesses that linger to this day. More than 2,000 animals were killed, and environmental damage was considerable. Bhopal was not equipped to handle an accident of this magnitude. Hospitals and dispensaries could not accomodate the flow of injured victims; likewise, government officials and registered mortuaries could not keep up with the certification and burial of the dead.

A HOT Analysis of the Causes

The Bhopal accident was caused by a complex set of interdependent Human, Organizational, and Technological factors. This HOT framework shows how antecedent conditions for the accident developed and how a complex set of interacting failures led to the disaster.

Human Factors

Human operators play a central role in industrial production facilities. An industrial plant is not just machinery: it is a complex socio-technical system that requires frequent human intervention. Operators control technological subsystems and coordinate interactions between subsystems. Managers supervise the operators and make decisions that directly affect operations.

Accidents usually involve errors made by both operators and managers. These errors are a function of the overall quality of plant personnel, which depends on a number of factors: employee morale, the number of people staffing each unit, the quality of training, and managers' experience. All of these factors played an important role in the Bhopal accident.

The working environment of the Bhopal plant tolerated negligence and a lack of safety consciousness among workers and managers. Employee morale was low because the plant was losing money and being considered for divestment. Compared to other divisions within the company, the Agricultural Products Division offered poor career prospects. Low employee morale, combined with ongoing labor-management conflicts, contributed to carelessness in operations. According to the plant's 1982 operational safety survey, basic rules were being ignored. For example, maintenance workers were signing permits they could not read and leaving particular work tasks without signing off. Employees were working in prohibited areas without the permits required by plant rules. And the fire-watch attendant was sometimes called away to perform other duties.

Low morale prompted the company's best employees to leave. In the four years prior to the accident, 80 percent of the workers trained in MIC technology in the United States had left the plant. The others remained because they were local residents reluctant to leave the city. Many had started side businesses to supplement their income and gain job satisfaction.

In this environment, unsafe practices and decisions involving breach of policy were common. Several of these contributed to the accident, including the following:

1. *The number of operators manning the MIC unit was cut in half between 1980 and 1984.* The entire work crew was cut from twelve to six, and the maintenance crew was reduced from three to one. The position of maintenance supervisor on the second and third shifts was eliminated, a move that directly contributed to the accident. The maintenance supervisor was responsible for ensuring that there was adequate preparation for the pipe flushing operation, and that the slip blind was installed to prevent water from entering the tank.

Staff reduction also eroded the human backup system, which contributed to plant safety. Human backups were critical because the plant depended on manual inspections, control, and checking of instrument readings. Moreover, emergency communications within the plant, and between the plant and outsiders, depended on messengers running between different locations.

2. *Operators had inadequate safety training.* Many of the operators lacked a sufficient understanding of safe operating procedures. The use of safety equipment such as helmets, gas masks, and protective clothing was erratic. In the past, there had been many minor accidents, fires, and MIC leakages. One worker had died and others were injured as they were physically inspecting the systems. Most importantly, workers had no training in dealing with emergencies.[18]

3. *Managers and plant workers had little information on the hazard potential of the plant, and there were no emergency plans.* Top managers at the plant acknowledged their ignorance about the fatally toxic nature of MIC. They could not provide local authorities with enough technical information to cope with the accident and prevent extensive damage. Because there were no established emergency procedures, workers reacted in ad hoc ways that exacerbated the accident, For example, during the accident, the operators turned off the alarm that would warn the neighboring community about the accident.

4. *When storage tank E610 failed to pressurize on October 21, 1984, managers did not investigate causes of the failure.* Managers simply abandoned the tank and allowed it to store MIC without positive pressure. This permitted small quantities of water and contaminants to react with MIC and form trimer (a plastic substance) in the pipe lines. It was this trimer buildup that necessitated the flushing of pipes on the night of December 2.

5. *Operators failed to put in the slip blind to prevent water from entering the storage tank during the flushing operation.* When no water was coming out the other end of the pipe, the production supervisor ordered the worker who was washing the pipes to resume the flushing operation and did not investigate where the water was going. This supervisor had been transferred to his job just one month before from a UCIL battery plant and had little experience in MIC technology and little knowledge of hazards in this plant.

The Union Carbide technical report hinted at another human cause of the accident. The report suggested, and Carbide lawyers later argued in court, that water was deliberately introduced into the tank by a saboteur. The company's videotaped message to its employees described how this might have happened. But the only evidence of sabotage they produced was a report that an uncorroborated story had appeared in an Indian newspaper. The story indicated that someone had seen a street poster in which a Sikh group claimed responsibility for the accident. Union Carbide later announced that the saboteur was an ex-employee, but UC did not provide any evidence.

But accident by sabotage was technologically improbable because the accident had involved *simultaneous* failures in design, technological subsystems, safety devices, managerial decisions, and operating procedures. More importantly, some of these failures occurred several weeks prior to the accident. To intentionally bring about the accident, saboteurs would have had to control operations of virtually the entire plant for several weeks.[19]

Organizational Factors

Two organizational factors typically generate preconditions for accidents. First is a set of strategic factors that create certain internal pressures on operations. Second is a set of operating policies and procedures that determine safety variables. Both played a role in Bhopal.

The first strategic factor was the relatively low importance of the Bhopal plant to Union Carbide. If a plant is strategically unimportant, it receives fewer resources and less managerial attention, which, in turn, usually make it less safe. Despite the best attempts to standardize safety provisions, corporations are rarely able to follow uniform safety standards in all locations. Each location has different regulatory constraints, operational needs, and technological capabilities. These differences are used to justify the differential allocation of resources.

The Bhopal plant was an unprofitable unit in an unimportant division of the corporation. UCIL was one of fifty international subsidiaries of Union Carbide. It represented less than 2 percent of the parent company's worldwide sales and less than 3 percent of its profits.

The Bhopal plant, one of thirteen UCIL manufacturing plants, had operated below 40-percent capacity for several years. The Agricultural Products Division, which administered the plant, had lost money for three years and had few prospects of becoming profitable in the foreseeable future. Because of its relative unimportance as a business unit in the corporate portfolio, the Bhopal plant did not receive top-management attention or support.

The second strategic factor was that the parent company made several important mistakes in setting up the plant. It was established in the face of a declining and increasingly competitive pesticides market. Even as it was being constructed, Union Carbide executives debated its economic viability. They evaluated several alternatives, including calling the project off during construction, but decided to proceed because it was too far along to abandon.

In June of 1979, Union Carbide's World Agricultural Products Team sought to justify the Bhopal plant's existence by expanding its role. They suggested that the plant could be made viable by exporting excess MIC, or it could serve as a base for making Carbofuran, another pesticide component, for markets in the Far East. However, the Union Carbide Agricultural Products Company, which oversaw the corporation's agricultural products all over the world, vetoed this plan in 1983, and Union Carbide Eastern also withdrew support for the plan. Instead, in February of 1984, James Law, chairman of Union Carbide Eastern, proposed selling the Bhopal plant, with the exception of the MIC unit. The World Agricultural Products Team and Union Carbide's top management endorsed this plan in July of 1984.[20] At the time of the accident, the Bhopal plant was for sale.

A third strategic factor was top-management discontinuity at the plant. Over the previous fifteen years, the plant had been run by eight different managers. Many of them came from nonchemical-industry backgrounds and had little or no experience dealing with hazardous technologies. Discontinuity in top management created frequent changes in internal systems and procedures, as well as uncertainty for operating managers.

In terms of operating policies and procedures, the presence or absence of corporate policies regulating safety variables, such as plant-maintenance schedules, backup-safety systems, safety training, emergency plans, and work practices play an important role in plant safety. Technology policies and manuals and safety, maintenance, and operating procedures at the Bhopal plant all were adapted from the parent company's documents. Day-to-day management was largely under the charge of local managers. One exception to this practice was the periodic operational safety survey, a safety "audit" conducted by staff from the parent company. The last such survey of the Bhopal plant was conducted in May of 1982. While it found "no situations involving imminent danger or requiring immediate correction," it identified ten areas of major concern and instructed the local management to remedy them.[21] Five of these problems eventually contributed to the accident:

1. The potential for release of toxic materials in the phosgene/ MIC unit and storage areas because of equipment failure, operating problems, or maintenance problems
2. A lack of fixed water-spray protection in several areas of the plant
3. The potential for contamination, excess pressure, or over-filling of the MIC storage tank
4. Deficiencies in safety valves and instrument-maintenance programs
5. Problems created by high personnel turnover at the plant, particularly in operations

The local management team developed action plans to solve some of these problems.[22] While some of the audit staff's concerns were addressed, most of the problems were chronic and needed continuous monitoring. Descriptions of the physically rundown condition of the plant at the time of the accident suggest that these chronic problems still existed.[23]

The plant had no contingency plans for dealing with major accidents. The need for such plans was not identified, either by the local management or by the experts from headquarters who conducted the periodic operational safety surveys. This, in turn, contributed to the general lack of understanding, both in the plant and in the community, about the lethal nature of MIC.

Technological Factors

Accidents in complex, tightly coupled, interactive technological systems are caused by multiple failures in design, equipment, supplies, and procedures. Typically, system accidents occur because of unanticipated interactions among multiple failures.[24] One component's failure triggers failures in other components or subsystems. Due to the high complexity and level of interaction among subsystems, designers and operators are unable to predict failures or their mutual interactions. In addition, tight coupling between subsystems reduces operators' ability to control damages and recover from accidents once they are set in motion.

Such failures include flaws in the design of the plant and its components, the use of defective or malfunctioning equipment, the use of contaminated or substandard supplies and raw materials, and the use of incorrect operating procedures. Each of these failures can be inadvertent and unanticipated; they can be deliberate; or they can be caused by negligence on the part of designers, consultants, manufacturers, operators, suppliers, and managers. The sources and causes of failures are critical, because they determine liability for damages.

The technological preconditions for a major accident were embedded in the design of the Bhopal plant, which allowed for bulk storage of MIC in large, underground tanks in a production environment that used manual noncomputerized control systems.[25] This type of storage design is more dangerous than the two available alternatives—small drum storage and no storage of MIC in a closed-cycle production system.[26]

Other technological causes of the accident were identified by Union Carbide's technical team and scientists from the Council of Scientific and Industrial Research, who analyzed the residue in storage tank E610.[27] They confirmed that several technical problems contributed to the accident:

1. *Tank E610 was not under positive nitrogen pressure for almost two months prior to the accident.* During this period, small quantities of contaminants, such as metallic impurities, entered the tank through the nitrogen line. These impurities acted as catalysts in the MIC-water reaction, thus increasing the speed of the reactions. Residue samples taken from several locations in tank E610 confirmed the presence of these impurities.[28]

2. *The water that entered the tank reacted with contaminated MIC in the presence of these metallic impurities.* This caused a self-accelerating trimerization reaction that generated heat in enormous quantities, leading to temperatures around 250° C (approaching 500° F). At these high temperatures, secondary chemical transformations occurred, leading to the lethal mixture of products that were spewed from the stack. Without these metallic impurities, 1,100 pounds of water would have been insufficient to cause such a violent reaction in such a short period of time.

3. *The fifty to ninety parts per million of sodium found in the tank residue suggests that water entered the tank through the relief-valve pipe and the process pipe.* Alkaline water from the scrubber's accumula-

tor backed into these pipes. (Several liters of alkaline water was drained from the pipes in May of 1985.) The backup of alkaline water through these lines increased the possibility that metal contaminants would enter the tank.

4. *For water to enter the tank through the pipes, it would have had to go through the blow-down valve or the safety valve.* The possibility of the blow-down valve malfunctioning was suggested by the absence of nitrogen pressure in the tank and by the fact that MIC was leaking from the relief-valve vent header on December 2.

These conditions were the result of failures in design, equipment, supplies, and operating procedures. Table 3–3 summarizes these failures.

There were many factors interacting with these failures that allowed impurities and water to enter the tank and then made it impossible for operators to control the accident after it had occurred. The design of the plant allowed for the storage of large quantities of MIC. Production and inventory decisions permitted the storage of contaminated MIC. A failure in the nitrogen pressure system that continued for two months allowed metallic impurities to enter and remain in the tank, thus creating conditions for water to react explosively with MIC.

Water entered the storage tank because of multiple failures in plant design (the jumper line between the two pipes), faulty equipment (the malfunctioning blow-down valve), and procedural errors (pipes were washed without slip blinds, and the flushing of pipes continued even though it was evident that the pipes were not safely blocked from the storage tank).

The water reacted with the MIC in a self-sustaining, exothermic chain reaction that caused the temperature and the pressure in the storage tank to rise rapidly. The temperature rise could not be controlled because of a failure in the operating system, which had earlier shut down the refrigeration system meant to cool the storage tanks.

Gases generated inside the storage tank could not be neutralized or prevented from escaping into the atmosphere because of another series of interacting failures. The scrubber was designed to deal with gases alone and not with a mixture of gases and liquids escaping the tank. The flare tower was down for maintenance, and the scrubber failed to operate. Water sprinklers could not throw water high enough to neutralize the escaping gases. Thus, toxic

Table 3–3. Technological Factors Contributing to the Accident.

Design
- Computerized early warning system and data logger were not incorporated into the design.
- Process design allowed for long-term storage of very large quantities of MIC in tanks.
- Water sprays were designed to reach only 12–15 m although gases from the flare tower were released at 33 m.
- Jumper line design modification connected relief-valve vent header and process-vent header, allowing water ingress into storage tank.
- Maximum allowable scrubber pressure was 15 psi while the rupture disk allowing gas to come into the scrubber was rated 40 psi.
- A single-stage, manual safety system was used in place of an electronically controlled four-stage backup safety system used in other similar plants.
- The design did not provide a backup system to divert escaping MIC to an effluent area for quick neutralization as in Bayer's MIC Plant.
- Manual system used for scrubber startup is generally less reliable than automatic systems.

Equipment
- New plant was built in 1981 but poorly maintained.
- Instruments and gauges were unreliable.
- Valves and pipes were rusted and leaking.
- Refrigeration unit (30 tn capacity) was too small and erratic to be effective in case of runaway reactions.
- Safety and operating manuals were in English, not easily readable by operators.

Supplies
- Highly toxic MIC and phosgene gases used as basic raw materials without knowledge of their effects.
- MIC in tank E610 was contaminated with a higher chloroform level than allowable.

Operating Procedures
- Failure to pressurize tank E610 with nitrogen was ignored repeatedly.
- Refrigeration unit was shut down several months before accident.

Table 3–3 continued.

Operating Procedures *continued*
- Flare tower and scrubber were simultaneously nonoperational while large inventories of MIC were in storage.
- Spare tank was not empty for operators to transfer MIC to it.
- Tank E610 was filled to 75 to 80% of capacity. Recommended capacity was 50%.
- Water flushing of pipes was reordered without investigating what was preventing water from coming out the other end.

---

gases simultaneously bypassed three safety devices as they were released into the atmosphere.

These human, organizational, and technological factors combined to bring the accident about. However, they did not, by themselves, cause the accident to escalate into the worst industrial crisis in history. It became a crisis because the environmental conditions outside the plant—the large number of people living near the plant and an inadequate physical and social infrastructure—combined with technological failures to create secondary effects that vastly increased damages from the accident.

The Environment in Bhopal

Bhopal is a textbook example of a rapidly developing city that sought—and obtained—sophisticated Western-style industrialization without making a commensurate investment in industrial infrastructure or rural development. By examining the roots of this industrialization, we can better understand why conditions outside the plant evolved the way they did.

Throughout most of its 1000-year history, Bhopal was ruled by a succession of feudal Moslem rulers, called nawabs, and Hindu rulers, called rajas.[29] In fact, the nawabs of Bhopal ruled the city under British protection from mid-nineteenth century until India's independence in 1947. This gave Bhopal a pronounced Moslem character in the midst of a primarily Hindu nation. Because the nawabs derived all of their income from land revenues and agriculture, they did nothing to develop or modernize the city. Even at the time of independence, transportation, communication, public health, and educational services were primitive.

After independence, Bhopal's population grew rapidly for two reasons. The first was its selection, in 1956, as capital of the state of Madhya Pradesh, one of the least industrialized states in India.[30] The other was its strategic location and natural resources. Bhopal is the most centrally located city in India. It had a good agricultural and forest base, and two large lakes that ensured a steady supply of water to the city.

The government quickly became the most important segment of the local economy. Indeed, government was omnipresent in Bhopal. It was the largest employer, the largest producer, and the largest consumer. Its powerful position was strengthened by the continuity of Congress party rule since India's independence in 1947.[31] By the 1980s, under the leadership of Arjun Singh, the party's power and influence reached unprecedented levels. More than 90 percent of India's productive industrial resources are controlled directly or indirectly by agencies of the city, state, and central (federal) governments. Virtually all service organizations are nationalized, including banks, insurance companies, postal and telephone systems, radio and television stations, energy production and distribution, and railways, airlines, intercity bus service, medical services, and education.

The city's transformation into an industrial center began when Bharat Heavy Electrical Ltd. (BHEL) was established about two miles outside the city boundary in 1959.[32] Employing more than 50,000 people, this gigantic, publicly owned manufacturer of electrical equipment almost overnight transformed Bhopal from a feudal fortress filled with government barracks into a large industrial city.

The state government encouraged the establishment of industries ancillary to BHEL, as well as other industrial plants, in order to generate employment. However, it was unable to engage in successful regional planning and urbanization policies. Bhopal's rapid urban growth was market-driven, haphazard, lopsided, and dictated by political expediency.[33]

Government planning efforts were outpaced by the unplanned expansion of the city. Two proposals for development of the city were formulated—the Capital Projects Development Plan of 1958–59 and the Interim Development Plan of 1962–63—but neither was fully executed. To make matters worse, urban industrialization was not integrated with rural development. Agricultural production in rural areas was stagnant, while the state's population was growing at a rate of more than 2 percent per year. These conditions forced the rural unemployed to seek work in urban areas—turning Bhopal into a rapidly growing urban area. Bhopal's population grew from 102,000 in 1961 to 385,000 in 1971 and to 670,000 in 1981— a growth rate almost three times the average for the state and for India as a whole.[34]

Migrants from rural areas were hardly equipped to deal with the difficulties of urban life. In 1971, almost two-thirds of the migrants were unemployed. Of these, half had not completed high school and 20 percent were totally illiterate.[35] Bhopal's rapidly rising population, coupled with high land and construction costs, caused a severe housing shortage in the city. Even though most of the migrants were poor, government efforts to build housing resulted, for the most part, in the construction of expensive dwellings. Unable to afford housing, many of the migrants became squatters, illegally occupying land and creating slums and shantytowns.

Most of these slums cropped up around industrial plants and other employment centers. Slum dwellers served as a pool of cheap labor for industry, construction, offices, and households seeking domestic help. By 1984, Bhopal had 156 of these slum colonies, home for nearly 20 percent of the city's population. Two of them— Jaya Prakash Nagar and Kenchi Chola—were located across the street from the Union Carbide plant, even though the area was not zoned for residential use.

Huts in the slums were small, unhygienic, temporary tenements, consisting of mud walls and wood, cloth, and tin-sheet roofs. They had no water or sewage services. Slum residents were often exploited by money lenders, protection racketeers, and illegal "landlords" who exercised control over the property by threat of violence.

Seeking to improve slum conditions and stop extortion, the state government passed a law in April of 1984 conferring ownership rights on the squatters. Each family was allowed to claim ownership to up to 50 square meters, or about 500 square feet, of land they were already occupying. But recognition of squatters' rights did not change the physical living condition of slums dwellers because the money needed to improve houses was not available.

The city's industrial infrastructure was no better than its housing. Although the area was not plagued by the electricity shortages endemic in many industrialized sections of India, the supply of energy was erratic. Voltage fluctuations and frequent power cuts necessitated the use of voltage stabilizers even for home appliances. In addition, public access to telephone service was extremely limited. While the city was served by 10,000 telephone lines, there were only thirty-seven public telephones. The remaining lines linked business and government facilities and a few privileged private residences. Even for those who did have access to phones, the unreliability of the telephone system was a source of constant frustration. Dead lines and barely audible connections were commonplace. Radio communication was limited to one government-operated radio station.

Although Bhopal was well served by rail, air, and incoming roads, the inner city (and the areas affected by the gas) had only a few, narrow streets. Large numbers of pedestrians, bicyclists, and animals on the streets controlled by only a few traffic lights contributed to congested traffic.

Bhopal had four major hospitals and several dispensaries and clinics that provided primary health care. But the city had a total of only 1,800 hospital beds and 300 doctors. (BHEL, which had its own township, also had its own hospital for employees.) Furthermore, the public health infrastructure was extremely weak, even for those lucky enough to live in permanent residences. The city provided water to residential taps for only a few hours a day. The tap water was of notoriously poor quality, and gastroenteritis was a chronic problem among residents. Poor water quality was traceable to the city's lack of raw-sewage-treatment facilities. Open sewage dumps were common in many parts of the city, and city sewage was indiscriminantly dumped into the two local lakes, one of which served as the city's main source of drinking water.

Despite its many shortcomings, residents considered the industrial infrastructure in Bhopal to be *better* than that of most towns and cities in Madhya Pradesh. As the state's capital, Bhopal was home to government administrative offices, the state assembly, important financial and industrial organizations, and many powerful politicians. Unlike other towns, it had attracted resources from both the public and private sectors.

The social infrastructure supporting industrialization had several shortcomings. India, as a whole, and Bhopal in particular, had extensive systems of laws governing industrial production, occupational health and safety, labor relations, trade practices, and pollution control. Bhopal possessed a large pool of skilled labor trained at local technical institutions. These attributes created a unique "industrial culture" that supported a wide range of technologies. But effective regulation of technologies was inhibited by political and practical considerations.

The government was reluctant to place a heavy industrial safety and pollution-control burden on industry for fear of losing job opportunities.[36] Industrial polluters often incurred fines so small ($40 to $400) as to be meaningless. Added to this lack of political backup was the paucity of financial resources and administrative capacity allocated to environmental protection. In 1982–83, the Department of Environment of the central government had only 150 employees (including only 25 scientists), a budget of $650,000, and a directive to concentrate on deforestation, the nation's primary environmental problem.[37] At the state level, control of industrial pollution, a leading cause of industrial crises, was handled by meagerly funded state agencies. Sophisticated industrial safety and pollution monitoring and control technologies were available but not widely used because of their cost.

India's thriving industrial culture was not evenly developed. It was limited to the modern urban sector of the economy and had many gaps that made industrial production hazardous. General citizen awareness of the hazards of industrial production was low. There were no community watchdog groups monitoring industrial hazards. Even warnings about the Union Carbide plant by a local journalist and some workers and earlier accident reports were ignored by local authorities and the public.[38] The struggle to survive and earn a basic living was the primary and, often, only concern of residents.

The accident at the plant was frightening because of the degree to which it involved human, organizational, and technological failures. But the devastation following the accident was also characterized by a similarly unpredictable interaction of events, people, and organizations—a sort of social and organizational equivalent of what occurred on December 2 and 3 inside the Bhopal plant. As we will discover, the devastation wrought by the crisis continues to mount even two years later.

CHAPTER 4

The Controversial Consequences
of the Bhopal Disaster

Even had Bhopal somehow recovered completely from the accident after the first day, the results would have been tragic. But the consequences of the accident stretched around the globe—to other parts of India, the United States, Europe, and South America—and have continued to exert their effect, month after month, year after year. Attempts to deal with the consequences have been marked not by sympathy and support, but by conflict and controversy.

It was these lingering consequences that transformed Bhopal into more than just a tragic accident. The death and destruction from the accident had secondary repercussions on victims' families and on the local economy. Madhya Pradesh remained in a state of crisis for several months. Union Carbide was severely damaged financially and as a corporate entity. And media headlines all over the world asked the question: "Can it happen here?"

Consequences in Bhopal

Immediately after the accident, while the immense government bureaucracy was still being mobilized for action, the major portion of relief was provided by volunteers from Bhopal and other parts of India—individuals, charitable institutions, social workers, and social clubs. They organized spontaneously, and they did virtually everything, including collecting and distributing food for the injured, transporting victims to the hospitals, providing first aid, counseling victims, providing shelter to orphans and destitutes, and arranging for the burial or cremation of corpses. More than forty volunteer organizations and hundreds of individuals provided social and medical services for several weeks after the accident.

Although the government bureaucracy was slow to mobilize at first, it did turn out a massive effort—95 percent of all the resources available for relief—that assisted thousands of people who otherwise might have starved or suffered more serious damage from their injuries. Because Bhopal was the state capital, the government was able to act with relative speed in providing relief. In fact, the state government established a separate Department of Relief and Rehabilitation for this purpose.

Simultaneously with the government's relief efforts, two sets of volunteer groups were formed to do relief work on a permanent basis. The first set consisted of social and charitable groups concerned primarily with medical and social rehabilitation.[1] The second included social activists interested in the larger medical, social, and political questions raised by the crisis.[2] These groups organized victims, raised their level of awareness, assisted them in filing for government relief benefits, and helped them to articulate their needs.

Over time, these groups built up a network of concerned citizens who protested slack management of the crisis by both the government and Union Carbide. Activists repeatedly came into conflict with Union Carbide and with state and central governments in India. Time after time, they criticized the government's relief efforts and frequently suggested that the government, as well as Union Carbide, had a vested interest in keeping the truth from the victims. These criticisms caused controversy in regard to virtually every aspect of the relief effort, in particular, assessing the exact number of deaths and the appropriate treatments for survivors.

The Death Toll Controversy

Remarkably, no one knows how many people actually died in the Bhopal disaster. A few months after the accident, the Indian government officially put the death toll at 1,754. But various sources suggest a wide range of higher figures, and the best conclusion one can draw is that the death toll was probably close to 3,000.

In estimating the death count at 1,754, the government included information from incomplete morgue records, then added figures reported from registered cremation and burial grounds and from out-of-town hospitals and cremation/burial grounds, eliminating

Table 4–1. Human Deaths and Injuries Caused by the UCIL Accident.

	Deaths	Injuries	Source of Data
Indian government	1,754	200,000	Law suit filed in N.Y.
Indian newspapers	2,500	200,000–300,000	*Times of India, India Today*
U.S. newspapers	over 2,000	200,000	*New York Times, Washington Post*
Voluntary organizations	3,000–10,000	300,000	
Delhi Science Forum	5,000	250,000	*Social Scientist*
Eye-witness interviews	6,000–15,000	300,000	Personal interviews[a]
Circumstantial evidence of death	10,000	–	• Shrouds sold • Cremation wood used • Missing persons estimated

a. Personal interviews with over 100 residents of Bhopal were conducted beginning on December 3, 1984. Even on the morning of December 3, local residents claimed that nearly 6,000 people had died.

previously accounted names. The media revised these figures upward, based on on-the-scene assessments by reporters.

Newspapers placed the total at between 2,000 and 2,500, while various other estimates—including those of social scientists, eyewitnesses, and voluntary organizations and figures gleaned from circumstantial evidence, such as shrouds sold and cremation wood used—ranged from 3,000 to 15,000 (see Table 4–1).

The problem of determining the exact number of deaths in such a situation is not new, and, in India, often becomes mired in political considerations. Mass deaths in religious and caste riots in India are routinely underreported by government sources in order to prevent conflicts from escalating. For example, in the 1984 Punjab riots involving Hindus and Sikhs, the government reported 2,000 deaths, while press reports and nongovernmental investigations put the figure at between 8,000 and 10,000.[3]

There were many reasons for discrepancies in death toll figures for the Bhopal accident. There was no systematic method to certify and accurately count the dead as they were discovered or brought to government hospitals and cremation or burial grounds. For the first three days after the accident, all available medical personnel were engaged in caring for the injured. Few people were left to care for the dead, register them, perform inquests and autopsies, issue death certificates, or arrange for systematic disposal of bodies.

Dead bodies piled up, one on top of another, in the only city morgue and in temporary tents set up outside of it. Many bodies were released to relatives for disposal without death certificates. Bodies were buried or cremated at unregistered facilities. Graves and funeral pyres registered as single burial units were made to accommodate many corpses because of worker and material shortages. Several people who ran from Bhopal died on roads outside the city and were buried or cremated by the roadside. These bodies were not registered with local authorities because accompanying travelers were too sick themselves or unable to make an identification. Within the city, unauthorized burials and cremations continued unchecked, and some unregistered deaths in neighboring districts were not counted for over a year.[4]

In the weeks that followed, the rate at which deaths and injuries continued overwhelmed the city's record-keeping system. In the first three weeks after the accident, more than 160,000 people were treated at the city's hospitals and almost 7,400 were admitted, even though there were fewer than 1,800 beds.[5]

The controversy over the death toll was linked to another issue that would arise throughout the crisis: the Indian government's dilemma with explaining its own role in the crisis while protecting its legitimacy as a competent and conscientious protector of public safety. While the government steadfastly insisted that only 1,754 people had died, personal interviews with senior government officials revealed that even they did not believe the government count to be accurate. Officially, they defended government figures as the most authentic figures available given the circumstances; privately they agreed that counting errors, if rectified, could easily raise the death count to around 3,000.[6]

Social activists assisting the victims were extremely skeptical of the official death count. They put forth a conspiracy theory, suggesting that the government and Union Carbide had intentionally reduced the official death count by secretly disposing of bodies en masse without registering them.

These activists argued that both Union Carbide and the government had a vested interest in disguising the truth with regard to the death toll. Union Carbide, they suggested, wanted to reduce the figure to minimize its liability from the accident. The government and the ruling Congress Party, according to the activists, wanted to control the potentially damaging effect of the accident on upcoming elections for the national parliament and state assembly. Because the government was perceived as being partially responsible for the tragedy by permitting slums to develop in the vicinity of the plant, a large death toll would mean a major political scandal.

Though little confirming evidence was available to support the conspiracy theory, it was repeated so frequently and by so many people that it began to gain credibility. The fact that both the government and Union Carbide had opened themselves up to this kind of criticism by not providing credible information was one of the first indications that the aftermath of the accident would be characterized by controversy.

The Controversy over Long-Term Medical Effects

At first, doctors in Bhopal believed that the aftereffects of exposure to MIC would be negligible or nonexistent.[7] They assumed, among other things, that MIC exposure had only immediate effects and that MIC was the only toxic agent involved in the accident.

But in the aftermath of the accident, this sanguine prognosis became less tenable. In the months that followed, it became clear that although MIC had been used in industrial processes for several decades and was known to be extremely hazardous, little was known about its long-term effects.[8]

Moreover, it was discovered that other toxic gases, in addition to MIC, had escaped from the storage tank during the accident and caused injuries. These gases were the by-products of the volatile chemical reactions in the tank.[9] Victims continued to suffer from breathlessness, dry cough, chest pains, restrictive lung diseases, dry eyes, photophobia, loss of appetite, nausea, vomiting, diarrhea, abdominal pain, menstrual disorders, and, in nursing mothers, suppression of lactation.[10]

The most serious and permanent damage among the injured was in the respiratory tract. Many victims died of oedema (fluid in the lungs). MIC also damaged mucus membranes, perforated tissue, inflamed the lungs, and caused secondary lung infections. Many survivors could not be employed because they suffered from bronchitis, pneumonia, asthma, and fibrosis, and were physically unable to work.

With the assistance of the Indian Council of Medical Research (ICMR) and the Gandhi Medical College in Bhopal, the government conducted several studies of the surviving victims. Interim morbidity data from ICMR studies are shown in Table 4–2. As is apparent from this table, the data released by the government were incomplete and unsystematically presented.

In August of 1985, the government issued a report entitled *Bhopal Gas Tragedy Relief and Rehabilitation – Current Status*. The report stated:

> The principal study being conducted relates to the large-scale and epidemiological aspects of MIC exposure. This is being done by the Gandhi Medical College in collaboration with the ICMR. Ten areas severely and moderately affected and three unaffected areas near Habibgunj have been chosen for the study. The exposed area sample consisted of 17,028 families with a population of 85,670 individuals, of whom 7,117 reported suffering from lung troubles, 6,993 from eye problems, 3,741 from GIT, and 2,516 from skin problems, with a total morbidity of 13.3%, while the control (unexposed) population consisting of 3,501 families and 15,632 individuals had 120 cases of lung trouble, 419 of eye problems, 31 of GIT, and 17 of skin problems with a total morbidity of 3.4%.[11]

These figures were inconsistent with other government reports, and were called into doubt by medical surveys done by nongovernment volunteer groups, which identified more severe problems, higher rates of morbidity, and further prolonged medical effects. They found the rates of disease among victims who lived near the plant to be significantly higher than rates among a control group that lived five miles away from the plant. For example, 85.4 percent of victims living near the plant were found to have moderate to severe illnesses: 94.6 percent had respiratory symptoms, 90.7 percent had eye problems, 53 percent had gastrointestinal prob-

Table 4–2. Morbidity in Affected Neighborhoods as of March 23, 1985.

Name of Locality	Number of Families	Total Population	Lung Cases	Eye Cases	Pregnant Women	Live Births
Heavily affected areas						
Jaya Prakash Nagar	1,104	5,475	1,607	1,800	121	32
Kazi Camp	1,235	6,861	314	256	253	49
Kainchi Chola	1,116	4,551	410	240	96	36
Railway Colony	1,105	5,748	340	237	65	45
Moderately affected areas						
Teela Jamalpura	1,078	4,028	266	200	54	20
Station Bajaria	1,007	5,059	500	424	93	28
Chand Bad	1,092	5,272	487	404	67	17
Straw Products	1,132	5,262	121	93	90	25
Slightly affected areas						
Bus Stand	1,242	5,606	1,307	1,249	132	3
Noor Mahal Area	1,160	5,890	1,013	982	57	21
Hawa Mahal Area	1,094	6,022	704	680	71	11
Shahajanabad	1,118	6,230	478	42	210	20
Fateh Garh	1,076	5,542	76	71	117	29
Not affected (control group) area						
Anna Nagar	978	4,482	54	54	81	24
BHEL	1,171	4,935	24	33	62	19
Habibganj	1,000	4,176	40	32	79	20
TOTAL	17,708	85,139	7,741	6,797	1,648	399

lems, and 43.6 percent had neuromuscular problems. These surveys estimated that out of 250,000 affected residents, about 60,000 suffered severe disability and about 40,000 more had a mild to moderate disability.[12]

Critics suggested that government study results were spurious because surveys were not conducted rigorously, accurate records were not maintained, the surveyed population was transient, and victims exaggerated damages to claim higher compensation. In any case, these studies did not lead to a clarification of the medical effects of the accident on victims. The government claimed that the results of the study were evaluated by medical advisory committees in order to establish better medical treatment protocols. In fact, however, no new protocols ever were implemented. In addition, the study and evaluation results were never released publicly. Therefore, their scientific validity could not be established by independent researchers. This, in turn, created confusion and contradictions about the nature and extent of physical damages caused by exposure to MIC.

The Cyanide Controversy

Just as no one knew for sure how many people died in Bhopal, no one knew exactly what leaked from the storage tank on the night of December 3. It was certain that the mixture contained MIC highly contaminated with chloroform. But reactions in the tank had a complex chemistry. Several intermediate compounds formed by the high-temperature decomposition of MIC escaped into the air along with MIC.[13]

Among the suspected compounds was hydrogen cyanide, which is a deadly poison, but one for which antidotes are known.[14] The same volunteer medical surveys that refuted the government's injury figures also found that 35 percent of the patients had contracted gastrointestinal, central nervous system, and eye problems, even in the absence of lung damage, thus raising the possibility that a cyanide-like toxin was circulating in the blood stream of many victims.[15]

Like questions concerning the death toll, the cyanide controversy, which raged for several months, carried significant political overtones. Volunteer social workers and political activists argued that it was in the interest of both Union Carbide and the government to underplay the role of cyanide in the accident.

Proof of cyanide poisoning would have completely changed the nature of the crisis. Union Carbide would have become liable for more serious damages, because cyanide, unlike MIC, was a well-understood poison. Operations involving cyanide required a known set of safety procedures different from those that existed at the Bhopal plant. In addition, proof of cyanide poisoning would have caused a major political scandal for the state and central governments because they failed to identify it immediately and administer sodium thiosulphate, the known antidote.

In the months following the accident, several pieces of evidence suggested that treatment with sodium thiosulphate would yield better results than the simple symptomatic remedies most victims were receiving. A study by ICMR, as well as the experience of many doctors, indicated that symptoms were ameliorated in many patients treated with the antidote. Inconclusive autopsy findings showed discoloration of organs, cherry-red blood, and oedema of the lung and brain—all known symptoms of cyanide poisoning.[16]

Further speculation was fueled by a telex message from Union Carbide's medical director in Institute, West Virginia, who recommended the use of sodium thiosulphate if cyanide poisoning was suspected. The most important evidence of cyanide presence was a government study indicating the presence of hydrogen cyanide in the vicinity of the storage tanks immediately after the accident.[17] This study was not made public until well over a year after the accident. In addition, medical experts suspected that MIC could also have created cyanide inside the body by combining with hemoglobin, or that it could have increased the cyanogenic pool inside the body, leading to a cyanide-like poisoning.[18]

Union Carbide refuted claims that cyanide was involved, insisting that it was impossible to produce cyanide either in the storage tanks or inside the body.[19] In addition, a powerful group of local doctors (who, according to media reports, were paid consultants to UCIL) claimed that in their experience, the use of sodium thiosulphate treatment did little to alleviate symptoms.

Activists took it upon themselves to supplement government-sponsored medical relief efforts by opening a small clinic to administer sodium thiosulphate. This clinic, called the Jan Swasthya Samiti (People's Health Committee) Clinic, was staffed by volunteer doctors from throughout India. The government supplied the clinic with injections and syringes. Sodium thiosulphate treatment at this clinic provided symptomatic relief to hundreds of patients. Additionally, many individuals who could afford to do so had themselves tested for cyanide and took the cyanide antidote as a precaution. However, the medical establishment in Bhopal did not adopt the use of this treatment on a large scale.

The cyanide controversy raged for several months, until the Jan Swasthya Samiti Clinic and others were raided by police on the eve of a large demonstration planned by activists in June of 1985.

When the demonstration, held the following day, degenerated into a violent encounter between protesters and the police, the activists' momentum was broken, and many returned to their home cities. The cyanide question was never resolved.[20]

Economic and Social Disruption

The accident did tremendous damage to the local economic and social structures. In addition to the shutdown of the UCIL plant, two mass evacuations—the first at the time of the accident, the second during a fear-ridden "scare" two weeks later—led to the closure of shops, commercial establishments, business and government offices, and schools and colleges. These closures, and labor scarcity resulting from death and injury, disrupted essential services and civil supplies. The establishments that remained open had few employees and few clients.

Estimates of business losses ranged from $8 million to $65 million.[21] The closure of the Union Carbide plant alone eliminated 650 permanent jobs and approximately the same number of temporary jobs—jobs that were particularly important to the local economy because Union Carbide paid high wages. The plant shutdown also dismantled a $25 million investment in the city, which had provided secondary employment to about 1,500 persons. State and local governments lost untold thousands of dollars in taxes. The city, the nation, and the entire developing world suffered a loss of business potential because the accident damaged their business image.[22]

To make matters worse, relief efforts following the accident distorted prices and the availability of goods. At one point, almost 50 percent of the city's population was receiving free grain from the government. This caused grain prices to decline and labor prices to increase abnormally.

Family economies in the slum colonies, which had been, at best, balanced precariously, were totally disrupted by the loss of income and the addition of financial burdens. Each family unit was large— five to fifteen members—and children commonly contributed to the family income. Many family incomes were reduced by the death or injury of family members, particularly adult men who typically provided the major source of income. The financial viability of family units depended critically on the nonearning members who often worked up to ten hours a day providing support services in the home—women doing domestic work, children collecting firewood or fetching water. After the accident, many women and children were incapacitated (or had died), forcing other members of their families to give up income-producing activities in order to do domestic work. Many families were reduced to beggary.

Many survivors were forced to join new family units. Orphans and widows lived with relatives and friends, and children left their parents' homes to become part of a relative's family. This reshuffling of family dependencies was socially disruptive and, often, financially unviable as well. Host families, frequently unable to provide for new family members and bear the cost of their medical treatment, were driven further into debt.[23] Most debts were owed to local loan sharks, known for their interest rates of 200 to 400 percent. To make matters worse, the debts carried by deceased persons were transferred to their widows, surviving children, or next of kin.

In many cases, the death of the adult males thrust women and minors into the role of head of their household. In an environment where women and children faced discrimination based on caste, sex, and age, these survivors lacked the ability to manage the family's financial affairs and pursue compensation. Thus, many of them became easy prey for exploitative middlemen and local moneylenders.[24]

Psychological Problems

Deaths, physical injuries, and economic and social disruption overshadowed another important consequence of the accident: severe psychological problems among victims. Six months passed before these psychological side effects began to be acknowledged by mental health researchers.

Fear, the primary psychological symptom, was evident immediately following the accident. Mass fear gripped the city, causing almost 200,000 residents to flee. Between December 16 and 19, while the remaining MIC was being neutralized, nearly 400,000 people fled the city in fear of another accident.[25] Many of these people pawned or sold all their belongings at distress prices, intending to abandon the city permanently.

Other psychological and emotional symptoms included sleeplessness, nightmares, anxiety, loss of libido, projection of guilt, increased family violence, and impairment of learning abilities in children.[26] These symptoms were further complicated by the religious and cultural symbols that some people used to interpret the meaning behind the accident. Some people viewed the accident as punishment from God. Others believed it was a curse brought upon families by specific family members. Some viewed the accident fatalistically. These beliefs led many victims to lapse into states of lethargy and inaction, to take punitive actions against family members, or simply to acquiesce. Many victims who had the physical ability to put their lives back together simply had no motivation to do so.

The psychological problems were most severe in women of childbearing age. A large percentage of the female population also suffered complex gynecological illnesses that confused them psychologically.[27] One study showed that women of reproductive age who lived in the worst affected neighborhood suffered from shortened and abnormal menstrual cycles. Fifty percent of breast-feeding mothers reported lactation failure.[28] An ICMR study revealed higher-than-average problems among pregnant women. Of 2,700 pregnancies, 400 ended in abortions and 52 in stillbirths. Of the live births, 132 babies survived only a short time, and 30 were born malformed.[29]

These problems caused considerable psychological trauma and social stigmatization of women within their families. In the absence of appropriate counseling, many women feared losing their reproductive capacity. They were afraid to report these ailments or be treated for them because they did not want their husbands and relatives to suspect that they would not be able to have children. The social responsibilities of women exacerbated their physical injuries. These women typically spent four to six hours a day preparing meals on wood- or coal-burning stoves (*chullas*) in poorly ventilated smoke-filled kitchens. As a result, they continually inhaled smoke, worsening their lung damage. The smoke also irritated their damaged eyes.[30]

Environmental Consequences

Damage to plant and animal life, while equally devastating, was not studied systematically because most of the available resources were deployed for mitigating human losses. The number of animal deaths probably exceeded 2,000 and included cows, buffaloes, goats, dogs, cats, and birds, although official government records put the figure at only 1,047. One 7,000 animals were given therapeutic care. Postmortems on farm animals suggested the possible presence of an undetected toxin, lending credence to the view that cyanide poisoning was involved.[31]

MIC exposure destroyed standing vegetation in surrounding areas. Of forty-eight plant species examined after the accident, thirty-five were affected to some degree, and thirteen appeared free from damage.[32] Over time even severely affected plants sprouted new leaves that were apparently healthy.[33] Most vegetables and wheat plants suffered dryness, scorching, bleaching, and similar symptoms. But fruits and other agricultural plants—including alfalfa, mango, mint, banana, and date palm—appeared undamaged.[34] Damage to soil and water was limited. A dramatic increase in free carbon dioxide and ammonia nitrogen was detected in the water shortly after the accident.

Although no long-term monitoring of plant and animal health was undertaken, the government announced it was safe to drink water from local sources and to eat vegetables grown locally so long as they were washed first. The public questioned this assertion. Because so little was known about the long-term effects of MIC exposure, how could the government know for sure that local water and vegetables were safe for consumption?

Consequences in the United States

Although the most tragic and immediate consequences of the accident occurred in Bhopal, the secondary effects quickly spread to the United States. The Bhopal tragedy was front-page news in the *New York Times* for two weeks. It sparked public debate throughout the United States about the role of American corporations worldwide and the safety of complex technologies. Union Carbide's MIC manufacturing plant in Institute, West Virginia, was shut down temporarily. American public interest and activist organizations formed coalitions to assist victims and examine the underlying causes of the accident. But most of the accident's consequences in the United States centered on two specific areas: the fate of Union Carbide itself, and the legal effort to obtain compensation for the victims, which eventually found its way into the American court system.

Consequences for Union Carbide

The accident threatened Union Carbide's very survival. In its aftermath, the company was subject to worldwide humiliation. The day after the accident, the Bhopal plant was shut down and local managers were arrested on criminal charges. When Union Carbide's chairman, Warren Anderson, and UCIL's top management rushed to Bhopal, they too were arrested. Company morale declined rapidly. A few UCIL employees were so ashamed of being associated with the company that they left voluntarily.

The company's reputation came under intense attack by the news media worldwide.[35] In general, Indian news reports were more critical of Union Carbide and the government of India than were news reports in the United States: more critical and accusatory; more questioning of the organizational, social, ethical, and moral aspects of the accident; more open in addressing conflicts and contradictions; and more descriptive of the plight of the injured. American news reports, by comparison, focused on technical and legal matters. They were more concerned about the possibility of a similar accident happening in the United States and the general issue of environmental pollution, and less critical of the company's behavior. In both countries, however, the public image of Union Carbide as a responsible company was seriously questioned.

The Bhopal accident triggered a series of sanctions and protests against Union Carbide all over the world. In India and the United States, public interest and activist groups initiated a variety of grass-roots campaigns against the company. In Breziers, France, where Union Carbide used MIC made in the United States to make pesticides, the local community objected to reopening the plant after it was shut down following the Bhopal accident. In Rio de Janeiro, Brazil, the state government decreed that MIC could not be produced, stored, or transported within the state. In Scotland, despite a local unemployment rate of 26 percent, the city of Livingston rejected Union Carbide's proposal to set up a plant to manufacture toxic gases.

During this period of scrutiny and backlash, several accidents that occurred at Union Carbide plants further deepened the company's crisis. Immediately after the Bhopal accident, Union Carbide had shut down its MIC manufacturing plant in Institute, West

Virginia. The company invested $5 million in additional safety features to prevent a similar accident and made public announcements that such a major accident could not occur in the United States.[36]

However, on March 28, 1985, the chemical mesityl oxide leaked from the Institute plant, sickening eight people in a nearby shopping mall. Then, on August 11, another chemical, aldicarb oxyme, leaked from a storage tank at the same plant, injuring 135 people, 31 of whom were admitted to local hospitals. Two days later, another leak occurred at a sister plant in Charleston, West Virginia. Although no one was injured, the leak was highly publicized and spawned further investigations into company operations.

Investigations revealed that twenty-eight major MIC gas leaks had occurred at the Institute plant during the five years preceding the Bhopal accident. One of them occurred just a month before the Bhopal leak, releasing 14,000 pounds of an MIC/chloroform mixture into the atmosphere.[37] These leaks were caused by failures in operating procedures, equipment malfunctions, and human errors—the same factors that contributed to the Bhopal accident.

The immediate effect of the crisis on Carbide's stock was predictable. From a pre-accident level of $48 a share, the stock dropped to a low of $32¾ within a few weeks. It struggled at between $30 and $40 for several months before rallying to $52 by the end of August 1985.

During this period, the company was under intense pressure on Wall Street. Standard and Poors dropped the company's debt rating to the lowest investment grade. There was renewed pressure on the company to execute the financial restructuring it had been planning for several years. Many of its stockholders lost faith in the company. In the year after the Bhopal accident, institutional ownership of Union Carbide stock dropped from 60 percent to 35 percent. It was estimated that by December of 1985, 30 percent of the company's stock had passed into the hands of takeover speculators.[38]

On August 14, the day after the second Institute plant leak, GAF Corporation announced what Wall Street had been suspecting for six months: that it had acquired 5.6 percent of Union Carbide's stock. Although GAF claimed the shares were acquired for "investment purposes," the company picked up another 1.5 percent within the next twelve days, clearly indicating that a takeover attempt would come soon.

Legal Consequences

Soon after the accident, lawyers from the United States arrived in Bhopal, formed partnerships with Indian lawyers, and started arranging to represent victims in multimillion-dollar personal injury lawsuits against Union Carbide. Within three months, 40 lawsuits had been filed in U.S. federal courts, 5 in U.S. state courts, and 1,200 in Indian courts.[39] The chronological development of the legal ramifications of the accident is shown in Table 4–3.

But Union Carbide was not the only party taken to court. Many victims also sued the government of India, charging it with negligence in allowing the disaster to occur. Some lawsuits pointed out the delays, incompetence, and corruption involved in the relief effort.[40] Others argued that the government was partly responsible because it had allowed Union Carbide to locate and operate the hazardous facility, and because it had legalized the slums around the plant early in 1984. Critics faulted the government for failing to act on the recommendation of its own Labor Department, which had urged a safety investigation at the plant, and for failing to prepare for the possibility of an emergency at the plant.

In March of 1985, the Indian government passed a law conferring on itself sweeping powers to represent victims in the lawsuit and manage all aspects of registering and processing claims. The following month it filed a lawsuit in the United States, charging Union Carbide with liability in the deaths of 1,700 persons, the personal injury of 200,000 more persons, and property damages. In addition, Union Carbide was sued by its own stockholders for not informing them of the risks involved in doing business abroad. Cases in the United States were consolidated under Judge John F.K. Keenan in U.S. District Court for the Southern District of New York in Manhattan. Compensation sought in the suits added up to well over $100 billion—ten times the net worth of the company.

Negotiations to reach an out-of-court settlement had begun even before the consolidation of the lawsuits. Such negotiations are common in complex cases involving mass injuries, such as the Agent Orange case, the Johns Manville asbestos case, and the Dalkon Shield case. A major reason is that the complexity of these cases precludes a quick trial and, thereby, denies the victims timely justice.

The negotiations occurred among Union Carbide's lawyers and an executive committee consisting of the government of India's

Table 4–3. Development in Law Suits Against Union Carbide.

December 9, 1984	First suit for $15 billion in damages filed in a West Virginia Court.
December 1984 and January 1985	Over forty-five suits filed against Carbide in various state and federal courts; 482 personal injury suits filed against UCIL in Bhopal; a $1 billion representative suit filed in Bhopal against UCIL and UCC; a suit in India's Supreme Court against UCIL and the government of India and Madhya Pradesh.
January 24, 1985	Judicial panel on multidistrict litigation consolidates 18 federal suits against UCC for pretrial proceedings, in the Federal Court of the Southern District of New York under Judge J.F.K. Keenan. Another 21 suits filed in Manhattan were consolidated later.
March 1985	The Bhopal Gas Leak Disaster (Processing of Claims) Ordinance, 1985 passed by Indian Parliament conferring on the Government of India powers to secure claims arising out of the disaster.
April 8, 1985	Government of India files *parens patriae* action against Union Carbide Corporation for personal injury and property damage claims.
April 25, 1985	Judge Keenan appoints an executive committee of plaintiff lawyers to manage the case. It includes Michael Ciresi, government of India lawyer; F.L. Bailey and S. Chesley, victim lawyers; and J. Hoffinger, the Liaison Attorney. Formation of the committee prevents the Indian government from completely dominating the case. Judge also asks UCC to provide $5 to $10 million for interim relief.
May 8, 1985	UCC offers $5 million for relief, to be deducted from payment of final settlement. It attaches stringent accounting requirements and demands detailed information on victims' health condition.
July 29, 1985	UCC moves to dismiss cases against it on *forum non conveniens* grounds. Judge Keenan asks for submission of a schedule for pretrial examination of witnesses, and allows limited discovery for resolving the forum issue.

Table 4–3 continued.

November 26, 1985	Court orders $5 million interim relief money to be given to Indian Red Cross.
December 6, 1985	*Amicus Curiae* brief filed by the Citizens Commission on Bhopal and the National Council of Churches arguing that the case should be heard in the United States.
January 3, 1986	Judge Keenan hears arguments from all parties on the *forum non conveniens* issue.
March 22, 1986	Union Carbide and private victim lawyers reach a "tentative settlement" of $350 million for compensation. Government of India is not party to this settlement and rejects it as absurdly low. The settlement is not implemented.
May 12, 1986	Judge Keenan rules on the forum issue sending the case to be tried in Indian courts.

lawyers and a group of victims' attorneys selected by Judge Keenan from among the private American tort lawyers representing Bhopal victims. Because the government had bestowed on itself all rights to represent the victims, it did not accept the role of private lawyers in the case. These lawyers had also lost legitimacy in the eyes of the victims and the world media because of the insensitive way they had descended upon Bhopal to sign up clients after the accident. They had obtained clients by running newspaper advertisements with affidavit forms attached, which the victims could fill out and mail back to the lawyers' respective offices. Some of them never even met their clients or discussed with them the nature or extent of the damages. These lawyers provided no follow-up guidance or advice, nor did they do independent research to establish facts about damages. Their main interest was in the extremely lucrative attorney fees that were likely to result from the case if it were decided in an American court.

All negotiators claimed that they sought a quick, fair, and equitable settlement for the victims. But they clearly represented conflicting interests and suffered from huge differentials in power that distorted communication among them. Judge Keenan attempted to balance the power of the opposing parties in order to keep them negotiating, but was not very successful. For example, in April of 1985, the court ordered Union Carbide to pay immediately $5 million for interim relief, deductible from the final settlement amount. But the government of India refused to accept the money, saying the corporation had imposed "onerous conditions" on its use. The court was unable to give away the money for seven months because the litigants could not agree on a plan for using it. This delay was embarrassing for all parties because, all the while, media reports detailed the woefully inadequate relief being provided to the victims.

Initial negotiations, which involved all three parties, led to Union Carbide's offer in August of 1985 of $200 million to be paid out over thirty years for a total and final settlement of the case.[41] The government rejected the offer without explanation.[42] Two detailed estimates of damage made public in 1985 suggested that the compensation to the victims should range from $1 billion to $2 billion (see Table 4–4).

In late March of 1986, the *New York Times* reported that a "tentative settlement" of $350 million had been reached between Union Carbide and the private attorneys. The agreement came shortly before Judge Keenan was expected to rule on the question of whether American courts were the proper forum for the case. Were the judge to rule that the case should be shifted to India, the private lawyers would lose their contingency fees. Thus, the lawyers had a strong economic motive for settling the case early, even if $350 million was too low to cover all damages.

But the Indian government's attorneys had not been involved in the negotiations, and they once again rejected the offer as absurdly low. Indeed, even if the agreement were sanctioned by the court, it would be virtually impossible to implement without the cooperation of the government, which was the only party with access to the information and administrative procedures needed to distribute the compensation money fairly.

In May of 1986, Judge Keenan ruled on the forum issue, deciding to send the case to India for trial. In doing so he imposed three conditions on Union Carbide. First, the corporation had to submit itself to the jurisdiction of Indian courts. Second, Carbide had to agree to satisfy any judgments rendered by Indian courts through due process. And third, the company had to agree to submit to discovery under the U.S. law, which allowed more exploration of

Table 4–4. Compensation Estimates.

Items	De Grazia (1985) U.S. $ Millions	Morehouse & Subramaniam (1986) U.S. $ Millions
Survivors of dead (3,000 persons @ $1,500 per work year)[a]	$127.50	$198.00
50% disabled persons (10,000 persons @ $750 per year)	198.75	
25% disabled persons (20,000 persons @ $325 per year	179.84	599.00
10% disabled persons (180,000 persons @ $150 per year)	720.00	
Business losses	64.00	8.00
Animal and property losses	1.00	1.00
Awards to helping groups	1.00	–
Cost of executing judgments/ settlements	20.00	–
Attorney fees	6.56	20.00
Statistical mapping of damages	–	10.00
Health care for victims	–	600.00
Food and nutritional supplements	–	105.00
Community infrastructure	–	120.00
Vocational rehabilitation	–	75.00
Epidemiological surveys	–	173.00
Future victims	–	134.00
TOTAL	$1,318.65	$2,043.00

a. Assumptions in parentheses are from A. De Grazia, *A Cloud Over Bhopal* (Bombay: Kalos Foundation, 1985).

company-held information than Indian laws did. This last condition was appealed by Union Carbide, which requested the court to make discovery under U.S. law a reciprocal condition and impose it on the government of India too. So nineteen months after the accident, the compensation issue was no closer to being resolved in the legal system. It was under appeal waiting to be transferred to India, where it would be taken up once again.

Consequences Around the World

Following the accident, business, government, and industry groups all over the world initiated responsive actions. Some of these responses marginally improved safety in hazardous facilities. Some led to new regulations imposed on industry in an ad hoc manner. Some merely expressed horror at the tragedy and Union Carbide's role in it.

Chemical companies around the world treated Bhopal as a technical accident that required incremental technical solutions. They individually reviewed their own safety operations and emergency response procedures, reduced the storage of some toxic chemicals, reevaluated the risks of operating in developing countries, and initiated programs for informing area residents about hazards.[43]

In the United States, the Chemical Manufacturers Association undertook two industry-wide safety initiatives—the Community Awareness and Emergency Response program (CAER) and the National Chemical Response and Information Center (NCRIC).

The CAER program was developed to provide public access to basic information on hazardous chemicals through material-safety data sheets, lists of workplace hazardous substances, and written hazard communication programs.[44] All of this information was already publicly available from different government sources, since it was revealed by employers under federal hazard communication standards.

The NCRIC program extended the existing Chemical Transportation Emergency Center, CMA's fourteen-year-old transportation emergency hotline. Four new services were added: a twenty-four-hour information line for emergency service and medical personnel available for accidents unrelated to transportation; CHEMNET, a mutual aid network of chemical industry and private emergency response teams to provide expert on-site assistance in emergencies; the Chemical Referral Center, to serve as a central contact point for information on chemical hazards; and development of training material for local emergency service personnel outside the industry.

Communities around the United States demanded more information on chemical hazards in their vicinity. Responding to these demands from the state and local levels, the Environmental Protection Agency (EPA) initiated the Chemical Emergency Preparedness Program (CEPP). The program provided information on the most hazardous chemicals (a list of 402 chemicals) produced in the country and guidelines for communities to develop emergency plans in cooperation with local authorities and industry.

In the United States, insurance companies sharply reduced coverage for toxic waste sites, while increasing the premiums. Since federal laws required toxic-waste operators to demonstrate financial responsibility, many dumpsites were threatened with closure.[45]

The Bhopal crisis also prompted legislative initiatives around the world. The United States Congress entertained several bills designed to improve the safety of manufacturing facilities and transportation of hazardous materials, as well as the dissemination of relevant information to communities.[46] Legislation was also passed in Belgium, West Germany, Great Britain, France, the Netherlands, and India. However, legislative initiatives in most countries were fragmented and addressed only immediate local concerns.

Environmental, consumer, labor, religious, and other public interest groups around the world took a keen interest in the Bhopal accident. Inspired by the volunteers and social activists in Bhopal who independently investigated the causes and consequences of the accident, citizen groups around the world initiated programs to monitor developments in Bhopal, help with relief work, and examine the implications of the crisis for communities everywhere.[47]

As these accounts suggest, the consequences of the accident mounted almost daily during the months after the accident and created an ongoing industrial crisis. But these consequences did not "just happen." They came about because the various stakeholders in the Bhopal accident, acting independently, took steps, which they thought were right, to cope with the situation. These steps then *interacted*, making the situation worse.

This is what typically occurs in an industrial crisis. To understand why the individual steps were taken, and how they interacted, we need to understand the frame of reference of each of the major stakeholders.

CHAPTER 5

Three Models of Crisis

A Multiple-Perspective Approach

Descriptions of what took place in Bhopal varied tremendously among the stakeholders. In its technical reports, Union Carbide referred to what happened as an "incident." The government of India, in its reports, called it an "accident." The injured victims called it a "disaster." And the social activists called it a "tragedy," a "massacre," and even "industrial genocide."

Nothing could reveal more starkly the differences in each stakeholder's frame of reference. In one sense, each description is correct. Yet each reveals an underlying, narrow frame of reference. Stakeholders' frames of reference dictated their goals as they responded to the accident, and hence their different, sometimes conflicting, courses of action.

To Union Carbide, the "incident" was a technical malfunction that needed to be corrected without causing major financial damage to the company. To the government, it was an "accident" that required relief without damaging the political position of the ruling regime. To the victims, it was a disaster that had irrevocably changed their lives; it required grief and anger and beginning the slow process of putting the pieces back together again. To the activists who sympathized with the victims, it was an unnecessary tragedy for which a negligent company and a culpable government ought to be taken to task.

If these stakeholders could have seen and understood each other's frames of reference, they might have been able to work together to achieve all their goals. But they did not. The stakeholders worked independently to achieve their own goals. While their actions each made sense from an individual point of view, when they interacted with each other they created a series of secondary effects that only served to deepen the crisis.

Perceptual differences among the stakeholders in any crisis arise because crises are, by definition, ill-structured situations and, thus, susceptible to many different interpretations. Furthermore, even in a crisis—perhaps *especially* in a crisis—self-interested behavior dominates. Each point of view serves the narrow interests of one stakeholder or another and suggests different solutions to crisis problems. To better cope with industrial accidents and prevent them from becoming crises, individual stakeholders must either change or somehow expand their frames of reference.

Stakeholders compete with each other to have their viewpoint accepted by the public as the "truth."[1] The establishment of one view as more "real" than the others is essentially a power game that involves adopting a set of partial solutions that benefit one stakeholder. To cut through this power game, it is necessary to achieve a multiple-perspective understanding of industrial crises.

Multiple-perspective analysis involves understanding and describing events from the perspective of all key stakeholder groups. This is done by acknowledging that social events are subject to multiple conflicting and disparate interests, assumptions, values, and interpretations, and then using those interests, values, and interpretations as a basis for building an understanding of events.

Examining individual crises from multiple perspectives has proven helpful in gaining a deeper understanding of other crises. An example is Graham Allison's examination of the Cuban missile crisis from three different perspectives within the government: the rational actor model, the bureaucratic process model, and the organizational politics model.[2] The rational actor model views crisis decisions as if they were made by President John F. Kennedy, who made rational choices from a series of alternatives presented to him by subordinates and technical experts. The bureaucratic-process model views those same decisions in terms of the interaction among bureaucratic organizations, such as the White House, the Pentagon, the State Department, the Department of Defense, and the National Security Council. The governmental politics model views those same decisions as a process of intergovernmental politics between the United States and the Soviet Union, mediated by the governments of allies on both sides.

Though it has many applications, multiple perspective analysis is particularly useful in examining the causes and consequences of industrial accidents. In researching the cases of the Santa Barbara oil spill and the Blueberry Reserve sour gas-well blowout, R.P. Gephart found that, as in the Bhopal case, corporations and the government attempted to minimize the extent of the disaster at all stages. Because their frame of reference was scientifically based, they appealed to scientific knowledge in describing the disastrous events, even though that scientific knowledge contradicted the personal experience of the victims. Following the accident they excluded the victims from discussions concerning the dangers posed by hazardous facilities. On points of conflict, the corporate view of reality prevailed and was used for developing solutions for preventing future disasters.

For example, hearings on offshore drilling at Santa Barbara were not open to the public; thus victims were cut off from debate over the preservation of their own environment. Victims did not fare well in legal challenges. Union Oil was fined only $500 for pollution for one day. The victims were also unsuccessful in other forums, such as Congress. As a result, the private interests of the oil companies dominated over public interests.[3]

A key analytical tool for understanding any crisis is the concept of frame of reference, which has been used extensively in research on the subject. A frame of reference is the method people or organizations use to select and process information. It reflects their biases, attitudes, and modes for making judgments. It is the lens through which an individual or organization views the world.

Frames of reference analysis is particularly valuable for understanding how and why organizations react to crises the way they do. Whether it is a government, a corporation, or a community group, every organization develops institutionalized procedures for processing information in response to changes in their external environment.[4] These procedures inevitably reflect the management's frame of reference.[5]

As time goes on, while the external environment continues to change, the typical organization becomes more dependent on its standard procedures. External changes are only partially noticed. As a result, management's frame of reference, even when it is no longer appropriate, is not challenged. Such organizations and managers suffer from what two recent researchers have called "unrealistic perception and...deficiencies of perceptual capacities."[6]

Thus, frame of reference is a critical determinant of whether, and how, an organization will respond to crisis and helps to explain why organizations often take actions that seem to the outside observer obviously wrong, tactless, or ill informed.

Achieving a multiple-perspective understanding of industrial crises requires understanding how various frames of reference differ. Each frame of reference can be broken down into a number of component parts, all of which have to do with the processing and filtering of information. These parts include:

1. *Data Elements.* These consist of the basic assumptions, concepts, or units of information that individuals or organizations use to construct reality. In other words, they represent information considered admissible for decisionmaking and reflect a bias toward certain kinds of information. Some organizations prefer to use only information that is quantified and objective. Others willingly use qualitative and descriptive information — "soft" information — from personal sources.

2. *Cognitive Maps.* Every person or organization has a particular way of arranging information into cause-effect relationships in order to make sense of that information and reach meaningful conclusions. Cognitive maps are conceptual schemes used to do this arranging, and, as such, they often guide organizations in defining and solving problems. These maps vary greatly. Some consist of a logically integrated set of casual relationships, while others are vaguely intuitive images of problems.

3. *Reality Tests.* This is the method by which persons or organizations validate the information they find, the inquiries they make, or the cognitive maps they create. This they do by finding and articulating a legitimizing connection between their information, inquiries, or cognitive maps and critical social and cultural experiences. Reality tests can be objective or subjective, or they may be rooted in traditions and customary practices.

4. *Domain of Inquiry and Articulation.* In addition to the three substantive components listed above, frames of reference also are characterized by two other elements: their domain of inquiry and their degree of articulation. The domain of inquiry delineates the boundaries of inquiry and concern, the relevance of particular variables, and the alternate frames of reference. For the most part, frames of reference are taken for granted. However, the extent to which they are articulated varies. Some may be expressed as assumptions that underly organizational inquiries. Large organizations sometimes articulate and codify their shared frames of reference in stated policies.

In Bhopal, as is the case in most industrial crises, the three primary stakeholders were the government, a private corporation, and the affected communities. Each stakeholder's perspective was incomplete and partisan, and each perceived, experienced, and reacted to the crisis differently. Because the Bhopal crisis was typical in this respect, it can be used to develop three models of the crisis, each with a different frame of reference, consequent responses, and policy implications. An understanding of all three models will enable us to achieve a multi-perspective understanding of the Bhopal crisis. The three frames of reference are summarized in Table 5-1.

The first model interprets the crisis from the government's frame of reference. Because they are large bureaucracies, governmental entities typically prefer to deal with objective, documented data rather than with anecdotal or subjective material. Similarly, this information is assessed using rigid and rational cognitive maps rather than intuition. Reality tests are embedded in the standard operating procedures of the government agencies.

In Bhopal, the government of India followed the first model closely. In reference to Bhopal, the term *government* refers to a group of agencies, including departments and public-sector undertakings of both the state government of Madhya Pradesh and the central government of India. Though these agencies had different purposes, resources, and activities, they served the same interests and were subject to political control by the same group of people.

The domain of inquiry was broad, because it necessarily included the relief effort, the protection of the public interest, and the legitimacy of the state. The government's frame of reference was well articulated and codified in its policies and procedures. The government had considerable experience in dealing with similar disasters and reacted with massive effort. However, because elected government officials controlled the bureaucracy and the technocracy, standard operating procedures were modified for political expediency. In particular, protecting the legitimacy of the government became an important goal and helped shape many of the government's decisions.

The second model interprets the crisis from the corporation's frame of reference. Corporate frames of reference are typically rigid and scientific, built from objective technical data and rational cognitive maps. In this way, they tend to legitimize the narrow

Table 5–1. Models of Industrial Crisis.

Dimension	Model I	Model II	Model III
Stakeholder	Government agencies	Corporations	Victims
Stakeholder identity	Government of India and Madhya Pradesh	Union Carbide Corporation	300,000 victims
Stakeholder structure	Hierarchically structured network of agencies under political control	Global, private corporation	Individuals, families, public interest groups
Stakeholder's Frame of Reference			
Data	Objective, social	Objective, technical	Subjective, social
Maps	Rigid, rational	Rigid, rational	Multiple, intuitive
Reality Test	Procedural	Scientific	Experiential
Domain	Political, social, relief	Technical, legal, financial	Medical, economic
Articulation	Well articulated	Partially articulated in policies	Low articulation
Antecedent conditions	Rapid industrialization and urbanization; Plant needed for jobs and product/services	Stringent regulation; Declining market and capacity underutilization	Rural to urban migration; Social displacement
Crisis environment	Weak infrastructure; Good relationship with UCIL	Competitive pressures to maintain profitability; Slums permitted in plant vicinity	Unawareness of hazard; Lack of safe housing

			Accident at UCIL
Triggering event	Failure in design, equipment, safety practices Union Carbide liable for damages	Accident caused by local mismanagement Sabotage	Accident at UCIL
Crisis processes	Managing immediate rescue and relief Medical relief and rehabilitation Legal aid to victims	Damage control by neutralization of MIC remains in the tank Chairman Anderson's arrest Precipitous fall in stock prices	Evacuation in panic Conflicts and struggle for aid
Crisis effects	1,773 deaths and 300,000 injured, uncertain effects Changes in key personnel	About 1,700 deaths, and 60,000 injured, no long-term effects Multi-billion dollar law suits Financial and market losses Deterioration of image	3,000 to 10,000 deaths, 300,000 injured Extensive delayed effects Further impoverishment
Responses	Medical relief and long-term rehabilitation Political control of crisis Compensation management Marginal regulatory changes	Public relations campaign Legal defense	Carbide and government sued Political self-organization and agitation

strategic thrust of corporations to grow economically and operate with technological efficiency, even when this means discarding other considerations such as environmental protection and safe operations.

In the case of a corporation such as Union Carbide—a "professionally" managed, technology-intensive multinational corporation—reality tests are also likely to be scientific and empirical. In Bhopal, this emphasis on objective information and rational decisionmaking narrowed the domain of Union Carbide's inquiry to the technological, legal, and financial concerns arising from the accident. That is why the corporation and its spokesmen sometimes seemed callous about the human suffering resulting from the accident.

The third model interprets the crisis from the victims' frame of reference. Whereas the frame of reference of the government and the corporation are not dissimilar—both are large, sophisticated organizations concerned with protecting themselves—individual victims and communities are typically unorganized and much less inclined to adopt rational processes. For the most part, victims use subjective data, usually conveyed by word of mouth. Their cognitive maps are intuitive, although in Bhopal the maps were strongly tied to religious and cultural traditions. Victims' reality testing is done primarily in terms of personal experience. Usually, the victims' domain of inquiry extends no further than their own economic and medical concerns.

In Bhopal, the victims lived a marginal existence. Not only were they disenfranchised from the political process, but because of their poverty they were even distanced from the mainstream of social life. As stated earlier, most residents thought the Bhopal plant manufactured some benign form of "plant medicine." Furthermore, they did not distinguish between UCIL, the Indian company, and Union Carbide Corporation, and equated the company's interests with those of the United States. This made it very difficult for the victims to understand the perspective of other stakeholders, much less act appropriately in response.

The three key stakeholders in the Bhopal crisis made decisions based on their respective frames of reference. In each case their actions served to resolve some aspects of the crisis and, at the same time, aggravate other aspects.

The Government of India

Both the state and central governments of India had considerable experience in handling industrial and natural disasters because they owned and operated most services and major industrial facilities as well as the railways, where accidents occurred most frequently.[7] The standard response to disasters by government agencies reflects the government's frame of reference. These responses include: (1) mobilizing available resources at the disaster site for damage control, (2) handling rescue and relief, and (3) preventing political repercussions.

Mobilizing Available Resources

The Indian government provided 90 percent of the rescue, relief, and rehabilitation efforts during the Bhopal crisis. These efforts included ex gratia financial payments, free food, and medical services. The financial payments were intended as temporary relief to tide victims and their families over during the crisis period. But these payments did not reach everyone. By the end of June 1985, the government, with the assistance of doctors who conducted medical examinations, had identified only about 1,000 "severely affected" persons and had given them an average of $118 each. About 14,000 "moderately affected" persons, also identified through medical examinations, had received payments of about $16 each. By the end of October, the survivors of about 1,499 deceased persons had received about $830 each.[8] In addition, 22,000 families with annual incomes of less than $500 received $125 each in "consumption grants," which were meant to cover immediate needs and were not seen as compensation for the accident.

In the first six months after the accident, the government distributed about $8 million in free food, for the most part grain and rice, to both affected and unaffected areas. By October 1985, this total had increased to $13 million, but the food distribution ceased by the end of the year.[9]

Medical relief efforts included the creation of a thirty-bed hospital, two clinics with X-ray and laboratory facilities, and seventeen dispensaries in the thirty affected municipal wards. These facilities treated about 3,000 patients every day. In addition, the Indian

Council of Medical Research (ICMR) conducted twenty-one research projects on the treatment of MIC-related ailments.

The government also initiated an economic rehabilitation scheme called STEP-UP (Special Training and Employment Program for the Urban Poor). In the first six months after the accident, this program gave seventy-nine persons loans of up to $1,000 each to start their own small businesses. In addition, the government helped Union Carbide workers find alternate employment in government enterprises after the plant was shut down.

Rescue and Relief Effort

Despite the massive resources plowed into relief efforts—massive at least by Indian standards—results were unsatisfactory. A year after the accident, the government declared thirty-six of Bhopal's fifty-six wards "gas-affected areas." These wards had a combined population of 300,000. Yet 85 percent of them had not received any financial assistance at all.

There were many reasons why the relief effort was not as effective as it should have been. One was a shortage of reliable information about the effects of the accident, which, in turn, hampered the design of the relief programs. In addition to the question of how many people had died or were injured, it was not even known exactly how many people lived in the affected areas. What little information was available remained under the tight control of the government. This raised anxieties among residents, politicized events, and prevented volunteer groups from being effective.

Undoubtedly, a certain amount of information failure was inevitable given the circumstances. But part of the problem also was that the government based its relief program on an ongoing medical-social survey of affected neighborhoods. The survey was not completed until several months after the disaster, and it was virtually impossible to identify, on the basis of the partial data, exactly who had been affected. As a result, relief efforts were not targeted to those affected by the accident. In fact, vocal, aggressive, and politically well-connected people received relief benefits more quickly and in larger quantities than did needier victims in less powerful neighborhoods. Likewise, the cyanide controversy also arose largely from a lack of information; doctors did not understand the effects of exposure to MIC well enough to determine whether cyanide treatment was warranted.

The government was unable to stop local moneylenders and middlemen from exploiting injured residents. These exploiters took commissions for their services in procuring relief benefits and confiscated relief money as repayment for earlier loans. These abuses prompted the government to substitute checks for cash in making relief payments. But the abuses persisted because many recipients did not have bank accounts, which forced them to cash their checks with the moneylenders at high discount rates.

The Handling of Political Repercussions

While the government was greatly concerned about the welfare of the victims, it was equally concerned about retaining its legitimacy and preventing political fallout. Repeatedly these two concerns came into conflict.

The question of the government's role in causing the crisis was an important one. The accident occurred in the midst of a political campaign, three weeks before national parliamentary and state legislative assembly elections. Negative publicity for the Congress party government posed a threat to its reelection. Critics questioned the government's historically close relationship with UCIL and some of the actions it had taken, such as legalizing the slums, that significantly increased the damage. Both the government and its parent Congress party needed to distance themselves as quickly as possible from the company.

Immediately following the accident, control of government activities passed to a small group of people that included key politicians and bureaucrats and was headed by Arjun Singh, who was chief minister of the state and located in Bhopal. This made mobilization of the massive bureaucracy relatively easy in the days and weeks after the accident. Later, control was decentralized to include officials from the central government, who were in charge of pursuing legal remedies and settlement negotiations in the United States.

The government tried hard to maintain its own legitimacy. Despite frequent media criticism of the government's role in the crisis, the Congress party won the elections. Through information control, the government was able to shift most of the blame to Union Carbide, even though governmental decisions and policies played a role in causing the crisis.

The government handled the political consequences of the crisis in two ways: by controlling information and by quickly identifying and punishing officials believed to be responsible.

Immediately after the accident, the government established complete control over most information, held almost all of it in secrecy, and then released it gradually in order to influence specific events. By doing so, the government was able to make social workers, activists, legal experts, and even Union Carbide itself dependent on it for information. This allowed the government to shape answers to critical questions—who was to blame, who should be punished, what should be done to prevent similar accidents in the future, how should the relief effort be organized, and how compensation for victims should be obtained—that otherwise could be turned against government agencies themselves.

The government's control of vital data took many forms. It impounded factory records. It sealed the premises off from outsiders. It prevented workers on duty at the plant from talking to the media or even to their superiors at Union Carbide. And it instituted its own judicial inquiry, which by its terms of reference made some independent inquiries into the accident less authoritative and others illegal. The government also sought to monopolize new information by refusing to share results of its socio-medical surveys, epidemiological studies, research on the long-term effects of MIC, and environmental impact studies.[10]

Those individuals who possessed firsthand knowledge of events were removed from the scene quickly. Within four months of the accident, key crisis administrators—people such as the mayor, the secretaries of several key departments, and Arjun Singh himself—had moved out of Bhopal. In addition, temporary crisis managers from the army and air force, medical personnel, and scientific investigators returned to their homes within a few months of the accident.

Finally, the government kept careful control over information collected through research and investigations. The results of medical research studies—and even their design—were not released to the public at the time of this writing, eighteen months after the accident. Some investigations, including the judicial inquiry, were simply discontinued before they presented their findings.

The government also worked effectively to identify and punish, at least in a symbolic way, those it felt were responsible. That is why three UCIL plant managers, including the works manager, were arrested immediately following the accident and why Union Carbide's chairman, Warren Anderson, and top UCIL officials were arrested upon their arrival in Bhopal three days later. These dramatic actions were taken to demonstrate that something was being done in response to the accident.

Responding to criticism about its own role in bringing the crisis about, the government identified and summarily punished several scapegoats. The state labor minister was asked to resign because he had not acted upon the recommendations of an earlier investigation into the accidental death of a worker at the UCIL plant. Several other officials from the Labor Department and the inspectorate of factories, who should have acted on the report, were asked to take a leave of absence pending an investigation into their actions.

Despite criticisms about the way it handled relief efforts, the government emerged from the aftermath of the crisis with its legitimacy intact. Although relief efforts did not meet the expectations of many people involved, they were heavily publicized by the government and did, in fact, provide far more relief than any comparable efforts on the part of Union Carbide. Union Carbide's negligible contribution to relief only enhanced and added credibility to the government's image in this regard.

The effectiveness of this political management became evident in the election results that followed the accident and in subsequent legislation regarding compensation claims. Although elections in Bhopal were postponed from the end of December 1984 to February 1985, opposition parties failed to make the crisis a political issue. The Congress party won all 239 seats it sought in the state legislative assembly, as well as 27 out of 35 seats in the national parliament from Madhya Pradesh.

Gaining legitimacy through relief efforts and making Union Carbide appear to be both responsible for the accident and unresponsive to relief needs made the government "morally" representative of victims' interests. This, in turn, facilitated the passage of the Bhopal Gas Disaster Bill in the national parliament, which gave the government the exclusive right to represent victims. With the government's legitimacy established, victims signed affidavits giving the government the right to represent them in recovering claims. The government sued Union Carbide in the U.S. federal court laying claims on the corporation's worldwide assets. Having established considerable leverage over Union Carbide, the government then engaged in out-of-court negotiations to reach a settlement.

Sincere efforts at discovering causes of the accident and providing relief were not as effective as they could have been because they were often shaped (or cut short) by political concerns. For example, the judicial inquiry commission set up to investigate the accident was disbanded before it gave its findings. Similarly, medical research studies were not made public even in spite of the fact that scientific controversies raged over the issue of cyanide poisoning. Had government officials been able to broaden their frame of reference enough to see the crisis in larger terms, they might have been able to minimize the aftershocks.

Union Carbide

As the owner and operator of the Bhopal plant, Union Carbide, unlike the government of India, could not distance itself from the accident. As a result, the corporation was far less successful at managing the crisis than was the Indian government. In fact, Union Carbide made some mistakes that embarrassed the company and made it a convenient scapegoat for all failures connected with the crisis.

This occurred, in part, because Carbide lacked control over vital information. But Union Carbide's rigid and narrow frame of reference, which was concerned primarily with the technical and financial consequences of the accident, also hurt the company by making it seem unsympathetic to victims and uninterested in the root causes of the accident.

Crisis management by Union Carbide occurred in two phases. In the first two or three weeks after the accident, the company focused most of its activities on controlling damage within the plant and neutralizing the remaining MIC. Subsequent activities emphasized managing legal issues, internal financial and managerial problems, and public image. The company succeeded in handling these problems from a short-term managerial perspective, but failed to prevent a further deepening of the crisis six months later.

Technical damage control was organized under highly stressful conditions because of the state government's hard-line response. The works manager and four key plant managers were placed under arrest. They were allowed to work within the plant, which was under the control of the Central Bureau of Investigations, but they were not allowed to communicate freely with top managers of UCIL and Union Carbide Corporation or with reporters. UCIL's top management did not receive support from state government officials and, of course, when Union Carbide Chairman Warren Anderson and two top UCIL officials arrived in Bhopal on December 7, they were also arrested.

Union Carbide identified three areas that needed immediate attention: neutralizing the MIC inside the plant, coordinating activities with government agencies, and handling legal affairs. The works manager served as the chief coordinator for these efforts, assisted by a group of three senior plant managers.

Within the plant, procedures to neutralize the remaining MIC, as well as to conduct an investigation into the causes, were initiated under the supervision of scientists from the Council of Scientific and Industrial Research. Normal shipments of incoming supplies were cancelled and essential supplies needed for neutralization and investigations ordered. Outside the plant, a system was set up to communicate with other managers and workers.[11] Six subfocal points were set up at different places in the city, where staff members came each day to sign a master roll, receive instructions, make reports, and exchange information. A subgroup also was set up to ensure that employees and suppliers were paid on time.

To enhance communication and boost the sinking morale of UCIL's staff in Bhopal and elsewhere, videotapes with messages from Warren Anderson, Union Carbide President Eric Flamm, and Vijay Gokhale, the managing director of UCIL, were shown to staff members in December of 1984 and during the following months. Industrial relations managers periodically met with workers and union members to apprise them of developments.

In contrast with this explicit effort to communicate internally, communications with external agents, such as the media, stockholders, and regulatory agencies were very guarded. The UCIL staff was instructed not to talk about the accident to reporters and other outsiders. This silence went on for six months, until Gokhale finally agreed to an interview with a reporter.[12]

A key failure during this period was UCIL's inability to establish a viable rescue and relief effort. Its immediate rescue efforts were limited to the work of one company doctor operating out of a small dispensary at the plant. Later UCIL provided emergency medical supplies and medicines worth $10,000 to local hospitals and donated $1.1 million to the Prime Minister's Relief Fund, a standing fund to assist in disaster relief efforts. But this money was used by and credited to relief work done by government agencies. Five months after the accident, Union Carbide, under order from Judge J.F.K. Keenan, was asked to give $5 million for interim relief. This money did not reach Bhopal for over one year.

Many of Union Carbide's actions backfired. Anderson's visit to Bhopal on December 7 was perceived not as an expression of genuine sympathy for the injured, but as an attempt to preempt lawsuits. Although the corporation offered to help government agencies in relief operations, the government rejected the offers without giving any official reason. (Unofficially, government officials said they did not want Union Carbide to exploit its donations for public relations purposes. They also wanted to establish the notion that the government could handle the relief effort on its own.) Union Carbide claimed that it offered $20 million to mitigate effects of the accident.[13] However, these offers required that recipients provide Union Carbide with information on the health of victims, which would then be used in Union Carbide's legal defense. The company also initially offered about $200 million as settlement for compensation suits. Because Union Carbide made the offer without any rigorous assessment of the actual costs of recovery, and because the $200 million was the amount for which the company was insured, this compensation offer was seen by many as an attempt to minimize financial loss to the company, rather than an effort to help victims. Both offers were viewed as insincere and generated a considerable amount of negative publicity about the company.[14]

After the first few weeks of the accident, crisis management activities shifted to Union Carbide's corporate headquarters in Danbury, Connecticut. This shift was prompted by the large number of lawsuits filed against the company in the United States, and by the inability of the UCIL management to conduct meaningful negotiations with the Indian government over compensation and relief assistance.

In court, Union Carbide's narrow and technically oriented frame of reference again proved harmful. The company's legal strategy was to defend itself aggressively against all charges by denying liability and to continue the battle in the courts for as long as possible. Carbide's lawyers tried to shift the blame from the parent company to UCIL by saying that local mismanagement and sabotage had caused the accident. The company also argued in court that the figures for damages should be low because, in terms of strict economics, the value of life in developing countries is low.[15] Although these arguments might have made sense within the company's highly rational frame of reference, they simply made Union Carbide look even worse.

Initially, Union Carbide stonewalled the press, communicating through guarded press releases. This strategy came about in part because the corporation lacked information about many crucial questions relating to the accident, and in part because it feared that the information it released would be misquoted or misused. Shortly after the accident, Union Carbide hired Burson Marsteller Inc. of New York, one of the largest public relations firms in the world, to help improve its public image. The public relations efforts concentrated on improving the company's image in the following areas: safety of its other operations, responsibility for the accident, corporation's financial ability to cope with the crisis, and corporate efforts to bring relief and compensation to victims.

Safety was Union Carbide's foremost public image problem. The company suspended all its MIC-related operations, conducted safety reviews in these operations, and invested $5 million in safety improvements at its MIC production plant in Institute, West Virginia. In press releases, press conferences, and congressional hearings, Union Carbide management repeatedly asserted the company's excellent safety record and insisted that an accident like the one in Bhopal could not happen at the Institute plant.[16]

In March of 1985, Union Carbide released its own internal report on the Bhopal accident. The report and a related press conference suggested that the accident was caused by sabotage, local operator errors, and local management lapses. The company accepted moral responsibility for the accident but not legal liability for damages.

Because of the lack of accurate data, it was impossible to estimate accurately the corporation's financial liability. However, in January of 1985, Chairman Warren Anderson announced that the company possessed enough financial resources to cope with all liabilities arising from the accident. He claimed that a settlement would soon be reached with the government of India that would resolve all claims against the company for about $200 million, the amount covered by insurance. This assessment was supported by sympathetic stock analysts, who spread the word that the crisis would soon be over for Carbide. Although Anderson's predictions did not materialize, they did persuade the company's auditors to give unqualified approval to Union Carbide's 1984 annual report. The company also protested when Standard and Poors lowered its debt rating. All these actions were designed to project the image of a financially robust firm.

To bolster its image with stockholders, key Union Carbide managers met with financial analysts and important stockholders to assure them of the financial soundness of the company.[17] The company also promised to implement the financial restructuring plans formulated before Bhopal. These plans included divesting poorly performing businesses and investing in new profit opportunities.[18]

The Bhopal crisis was a natural rallying point for consumer, environmental, labor, public interest, and church groups, all of which were highly critical of Union Carbide's role in the disaster and its lack of participation in mitigating its effects. The company hired communication and strategy consultants to help it communicate its viewpoint to these groups. It also provided information on a variety of controversial issues to reporters, independent researchers, Wall Street analysts, and other company watchers. It encouraged public expressions of support for its policies through "We Love Carbide" rallies held in Institute, West Virginia.

But the crisis inside the company proved almost as difficult to manage as that outside. The moral burden of thousands of deaths and injuries ravaged Union Carbide's strong corporate identity. Employees were demoralized and needed constant explanations and reassurance from the company. Simply maintaining the morale of the worldwide corporate work force and supplying enough information to handle local queries proved to be a key internal management task.

Internal communications were standardized and routed through the communications department. This enabled subsidiary and field units to respond to local public and media queries in a consistent manner. Two sets of videotapes were used to communicate top management's understanding of the crisis, their plans for resolving it, and its consequences for employees. The tapes painted an optimistic picture of events, assuring workers that things were under control and encouraging them to have faith in the company. The technical aspects of the accident were communicated to workers through a report from a Union Carbide investigating team, which became the official Carbide explanation of the accident.

As the crisis went on, however, Union Carbide's financial problems deepened. In December 1985, GAF Corporation announced its plans to acquire UCC (see Table 5–2 for a chronology of events). Earlier, UCC had bought back 10 million of its 70 million outstanding shares, at a cost of about $500 million. Some $500 million in employee pension funds were also reverted to the corporation.

To avoid takeover, UCC was forced to take a series of cost-cutting measures to save $300 million in pretax earnings. Staff in the United States was reduced by more than 4,000 workers. Some plants were closed and weak business units in alloy metals and chemical lines, worth some $500 million, were divested. A new environmental protection program was established at a cost of $100 million. And some assets were written off or written down, resulting in a one-time charge of $90 million in inventory, $675 million in fixed assets, and $100 million in plant closure expenses.

Probably the most demoralizing move was Carbide's decision to sell its most profitable division, the Consumer Products Division, which made batteries, antifreeze, Glad Bags, and other well-known consumer products. The company realized $2.2 billion from the sale and was able to retire the $1.1 billion debt incurred in warding off the takeover. But the sale reduced the size of Union Carbide by 20 percent.

Like the government of India, Union Carbide was hampered in its response to the crisis by its own narrow frame of reference. Like many other multinational corporations that deal with complex technologies, Carbide made many of its decisions—both before and after the accident—based on narrow technical, financial, and legal considerations. Had Carbide officials been able to view crisis from a broader, social perspective, they might have done more to accept culpability immediately, pay the "price" of the tragedy, and

Table 5–2. Union Carbide Takeover Struggle.

August 14, 1985	GAF Corporation announces it has acquired 5.6% of Union Carbide stocks for 'investment purposes'.
August 28, 1985	Union Carbide announces a plan to buy back stock, reduce work force, and restructure management, as a means of thwarting any potential takeover attempt.
August 30, 1985	GAF announces it has acquired 9.9% of UCC stock.
September to November 1985	GAF organizes money to attempt a takeover of UCC.
December 9, 1985	GAF offers to purchase UCC for $4.3 billion, at $681-per share in cash for most stock and securities for the rest.
December 12, 1985	GAF sweetens its offer to include all cash payment for all shares. It announces plans to sell half of UCC if its takeover bid succeeds.
December 15, 1985	UCC rejects GAF offer and proposes to repurchase 35 percent of its stock for $20 cash plus $65 in notes. GAF sues UCC to prevent it from buying back its shares with pension funds.
December 26, 1985	GAF raises its offer to $74 cash. It also advises UCC shareholders to accept Carbide's previous offer and then sell the package of cash and notes to GAF for $74.
January 2, 1986	GAF announces it is willing to enter into an acquisition agreement with Carbide for $78 a share.
	Union Carbide announces a 7-point defense plan, involving sale of its profitable consumer products division (20% of sales) for about $2 billion. Proceeds would be used to retire $1 billion debt and pay $28 in special dividend. It would also buy back 55% of its shares, split its stock, and raise its regular dividend.
January 9, 1986	GAF withdraws from attempt to take over Carbide, netting $80 million in profits.

engage in a sincere effort to assist victims. The company might even have escaped with its image intact, as Johnson & Johnson did by promptly protecting consumers in the Tylenol poisoning crisis.

Union Carbide's actions remained wedded to the narrow strategic thrust of any company's overall business interests, which, for the most part, emphasized financial and production goals at the expense of social concerns. Although Carbide officials were individually horrified by the tragedy, the narrow organizational frame of reference of the company prevented it from exhibiting a deeper social concern.

The Victims

The frame of reference of the victims differed radically from that of the other stakeholders, but it was just as narrow. It was subjective and intuitive and based on personal experience and knowledge, which, in most cases, was limited. The victims' view of the world was reflected in their attitudes toward their government. Although they were represented on local, state, and national bodies through periodic elections, the operative political control of the neighborhoods near the plant was accomplished through a network of strongmen. Many victims thus viewed their elected government and its bureaucratic machinery paradoxically as both an oppressive force in itself and as a protector of their interests from other oppressive forces in society.

This frame of reference led some victims into fear, cynicism, and an opportunistic attitude toward relief that only made the suffering of their neighbors worse. Others contributed positively to the relief effort, but felt a bitterness toward the government and Union Carbide that, while justified, ultimately inhibited the government and Union Carbide's ability to improve the situation.

Many of the victims felt the need to vent their anger and anguish, perhaps only to convince themselves that they had personally done something to ameliorate the situation. They also wanted to participate in the social and legal battles emerging from the accident. But they did not have the political power or the resources to enter into these battles, which were dominated by the government, Union Carbide, lawyers, and analysts.

In an attempt to become part of this discourse, in 1985 these victims engaged in several peaceful *morchas*, or demonstrations.

They also used vernacular poetry to express their sentiments. The first march, which took place on January 3, 1985, was organized by social activists who had come to Bhopal to provide relief services to victims. Ten thousand victims and their supporters wore badges saying *Dhikkar Divas*, a Hindi phrase meaning Denouncement Day. The march was intended to denounce those responsible for the accident and for the inadequate relief efforts.

Carrying placards and banners with verses and slogans, the victims marched from the Union Carbide plant to the residence of Chief Minister Arjun Singh.

The strong feelings that marked victims' frame of reference were apparent in their songs. Most of the songs, three of which are translated from Hindi to English here, were cast in a question-and-answer format. A few were explanations of events.

For example, one verse went like this:

Carbide Ne Kya Kiya?
Hazaron Ko Mar Diya,
Sarkar Ne Kya Kiya?
Hathyaron Ka Sath Diya.

Translated into English, the verse reads:

What did Carbide do?
It murdered thousands of people,
What did the government do?
It aided the murderers.

The most important sentiment in this verse was a categorical condemnation of both Carbide and the government as partners in crime. The victims' bitterness was directed at the government as well as the corporation because of its cozy relationship with Union Carbide, its willingness to allow manufacture of a lethal gas next to a residential neighborhood, and its lax implementation of safety policies.

But the government of India was not the only object of the victims' anger. The United States, as it was represented in India by Union Carbide, was also to blame, as this verse suggests:

It was the conspiracy of the dollar
That mixed poisons in Bhopal—
Made the poison in America
And dumped it in Bhopal.

Because most of the victims were uneducated and many were illiterate, they did not distinguish Union Carbide from its country of origin. Their anger with Union Carbide was thus projected onto America as a nation. In fact, the author, who was born in Bhopal and lived there for twenty-three years, was occasionally viewed with suspicion by the victims because of his affiliation with an American university.

Those affected by the accident believed they had been victimized by people that they should have been able to trust, as this next verse reveals:

What did doctors do?
They did not give us milk.
What did lawyers do?
They used us as pawns.

In India doctors and lawyers are highly respected members of the community, regarded by all as guardians of people's health and rights. The reference to milk is an allusion to the fact that because of disorganization and some distribution malpractices, some victims did not receive free milk as part of the relief efforts. The lawyers referred to are, of course, the American personal injury lawyers and their Indian associates who descended so insensitively upon the victims in the days immediately following the accident.

The January *morcha* was not the end of the victims' public activities. A second demonstration was organized on March 13 with the support of local journalists, writers, and trade unions. But the victims, and particularly activists who served as their spokespersons, were constantly in conflict with the government.

This friction reached its climax in June of 1985. In the middle of that month, a few patients at the Jan Swasthya Samiti Clinic suffered mild reactions to sodium thiosulphate, the cyanide antidote. Volunteer doctors stopped dispensing the drug and closed down their clinic. However, they decided to step up their protest efforts. They planned a large demonstration at the state secretariat on June 25.

Protesters were planning to make several demands. First, they asked that interim relief payments of Rs 1,500 (about $135) be granted to all gas victims, not just those earning less than Rs 500 per month. Second, they wanted sodium thiosulphate injections administered on a large and rapid scale. Third, they wanted an investigation into "the corruption and arbitrariness of government officials in charge of compensation and relief work." Fourth, they wanted the Bhopal plant to be converted into a nonhazardous industrial plant to provide employment for UCIL's former employees. And finally, they demanded that "deterrent punitive action" be taken against Union Carbide.[19]

The night before the demonstration was to take place, however, police raided the clinic and the homes of activists, arresting fifty people, including six doctors and many neighborhood residents whom the government had identified as agitation leaders. The next day, the government imposed Section 144, an emergency regulation restricting public assembly, in order to break up the protest rally. But the demonstration was not called off. Instead, demonstrators avoided the areas under Section 144 regulation. They reached the secretariat where they intended to present their demands to Arjun Singh in person, by traveling other routes.

For whatever reason, the chief minister did not meet with them. The demonstrators waited in a cordoned-off area while their representatives met with other government officials. But someone in the crowd broke the cordon and threw stones at police officers who were holding the demonstrators at bay. This incited a *lathi* (billy club) charge by the police, who beat up and arrested many demonstrators, including women and children. The police also closed down the Jan Swasthya Clinic.

This show of power by the government broke the confidence of the victims' movement. Many out-of-town activists and volunteer doctors returned to their home cities shortly thereafter. On the first anniversary of the disaster in December of 1985, a rally was held at the Union Carbide factory gates, at which 4,000 demonstrators burned effigies of Warren Anderson and government officials. But, by and large, the activists' struggle continued, only at a reduced level. They were limited by their meager resources and frustrated by their inability to influence the government's actions.

Though victims' responses were limited by their frame of reference to anger and bitterness, activists helped channel those feelings into demonstrations that forced the government and Union Carbide to become more accountable. And while the victims' movement was ultimately a frustrating (and inconclusive) experience, it reminds us that the community is the most crucial stakeholder in efforts to prevent and cope with industrial crises. As we shall see in the next chapter, large organizations, like the government and corporations, are not always motivated to change or expand their frames of reference, even when it is in their best interests to do so. Community-based organizations may sometimes be the only groups with the power to force such changes.

After this examination, we can see that each of the three stakeholders took the actions to deal with the post-accident problems that they perceived as important. However, none of the stakeholders had a multi-perspective understanding of the problem. As we stated before, it is difficult for stakeholders to adopt one another's perspectives because crises are ill-structured situations. Information is scarce, ambiguous, and, often, carefully orchestrated by stakeholders who possess it.

It is possible, however, to take steps to overcome this lack of multi-perspective understanding. These steps involve changes in underlying perspectives by expanding each stakeholder's frame of reference and freer sharing of information, both in advance of and during a crisis. It is to these steps that we now turn.

CHAPTER 6

Preventing and Coping
with Industrial Crises

Had Union Carbide and the government, showed more foresight, the Bhopal accident might never have occurred. Had the government and the community been more aware of the plant's hazards or known how to cope with an emergency, the accident might not have caused so many deaths. And even if deaths and injuries were inevitable, Bhopal would not have become a worldwide crisis if the different parties involved had worked together to alleviate after-effects. Industrial crises can be prevented, or, at the very least, coped with effectively, provided the different stakeholders adopt a broad multi-perspective view of crises.

Any set of solutions will necessarily be incomplete in addressing the multiple problems connected with industrial crises. In fact, in thinking about industrial crises, a concept such as solutions is not even useful. This is because crisis problems are ill structured, technically complex, and extend beyond technical issues to social, political, and cultural issues. The recommendations made in this chapter are not solutions in the strict, mathematical sense of the term. They are simply a group of questions and strategy issues that need to be addressed by various stakeholders.

The improvement of preventive and coping responses in relation to hazardous industries involves three steps: (1) the different stakeholders must acquire the will to change, (2) they must participate in alternative methods of resolving disputes and compensating victims, and (3) they must each take actions individually to prevent serious industrial accidents and cope more capably with those that do occur.

The Will to Change

After the Bhopal tragedy, it may seem impossible to imagine that corporations, governments, and affected communities will find the will to expand their frames of reference. But in each case, there are factors that are beginning to *compel* them to change.

Corporations feel some pressure to change. In Bhopal, a single incident severely damaged one of the largest corporations in the world. Furthermore, public opinion polls show a steady erosion in public confidence in corporations, largely because of industrial crises. Corporations *must* respond in order to retain their legitimacy and survive as institutions.

Accepting the idea that corporations should change their frames of reference requires a long-range view of survival. But the arguments supporting such a change are powerful indeed. First, of course, is a moral argument. Corporations responsible for industrial crises bear the moral burden of hundreds of deaths and environmental destruction. Such death and destruction is unacceptable under any code of morality. Just as important, is the ethical question raised by the size of multinational corporations. Because they are often larger and more powerful than the governments of the developing countries in which they operate, they bear an almost government-like responsibility for the well-being of the people who live near their plants.

The second argument is an economic one. The costs of crises are very high and they increase when corporations try to deny or ignore problems. Corporations are liable for damages caused by crises. In the worst cases this liability can lead to bankruptcy, as occurred with Johns Manville, which was overwhelmed by liability claims for work-related asbestos diseases. Cleanup costs are another problem. In the case of complex technologies such as nuclear power, these costs can run into billions of dollars.

Even if a corporation can bear the cost of liability and cleanup, it might not be able to survive on Wall Street. Any major accident is likely to erode a large corporation's base of stable institutional investors. Although stock prices typically rebound after crises, institutional investors lose confidence in management and "sell out" to speculators and arbitragers. This, in turn, may expose the company to takeover attempts that, even if fended off, can threaten corporate bankruptcy, as was the case with Union Carbide.

The question of compelling a change in a government's frame of reference is a difficult one. Certainly one lesson that emerges from Bhopal is that a government can protect its legitimacy without altering its frame of reference, provided it can control information and distance itself from the company involved. The Indian government did not completely succeed in this respect. Although the Congress party won the elections after the accident, and the government of India succeeded in setting the agenda for postaccident issues, it did so at the cost of some erosion of its credibility and a great deal of criticism, both domestically and abroad.

A more compelling reason for governmental change may be the foreign policy implications of industrial crises. As the scale of industrial crises grows, they are more likely to have international effects. This was exemplified by the fire at the Soviet Union's nuclear power plant in Chernobyl. Radiation from the accident affected over a dozen countries. It elicited severe criticism and sanctions against the Soviet Union from foreign governments. Such international repercussions will force governments to be more open and responsive in preventing and coping with industrial crises.

But stimulus for changing corporate and government frames of reference may come from the third major stakeholder—the individual communities located near industrial plants. Ordinary citizens facing technological risks can play a key role in forcing change. Grassroots movements for environmental, consumer, and public protection have helped establish and legitimize viewpoints that productively challenge government and corporate perspectives.

Our social system neither provides government the power to control corporate behavior with regard to crises, nor does it give corporations the incentives to prevent and cope voluntarily with crises. Government and businesses are critically dependent on each other in the sense that they both deal with the contradictory demands of increasing the productivity of private capital while simultaneously reducing its negative side effects on the public. This balance of corporate and government power perpetuates the status quo. Only the direct involvement of the public in policymaking and in monitoring implementation of crisis-related measures will structurally change this situation.

The resources required for preventing and coping with crises will have to come from both corporate and government sources. More investments in safety systems, insurance coverage for liability, injury compensation systems, and development of adequate infrastructures will be needed. Top management and governmental policymakers must be committed to making these additional resources available.

But the public can play a large role in this process by insisting on public monitoring of hazards within their communities and by putting appropriate pressure on decisionmakers. A healthy tension between corporations, government, and the public is essential for the progressive resolution of industrial crises.

We have seen effective intervention by citizens in some countries. In the United States, citizen and environmental groups have played a significant role in creating relatively high standards for environmental protection and industrial safety. In West Germany, the Green Party has become a national political force by effectively building on just these sorts of issues.

In developing countries public involvement is a more complex issue. The disparities of power between the educated ruling elite and the uneducated masses are so large that they preclude any effective participation by the masses in social control over technology. The daily struggle for survival prevents individuals from seeking solutions to longer-term problems of pollution and safety. Mass unemployment forces them to unquestioningly accept industrial hazards. As Bhopal so tragically demonstrates, the price of safety is eternal vigilance. Communities and individuals ignore this lesson at their own peril.

Alternative Ways of Resolving Disputes

The delay in compensating victims of Bhopal was the most scandalous and inhuman aspect of the entire disaster. The victims were poor and powerless, and the aftereffects of MIC exposure prevented them from earning necessary wages. And at the same time that the government suspended its free food distribution, a drought hit the central part of India, driving food prices up. As a result, unemployed and unemployable victims were driven to virtual beggary. They watched helplessly as Union Carbide, the government of India, and their private lawyers fought for their respective points of view in foreign courts and in the media.

The irony of the situation was that all the contestants righteously insisted that they had the victims' best interests in mind and wanted them to receive quick and equitable compensation. Their actions, however, were not consistent with their pious rhetoric.

Warren Anderson, the chairman of Union Carbide, claimed "moral responsibility" for the accident but could not find a way of helping with relief efforts. The company continued to "offer" and "pledge" up to $20 million in aid, but less than 10 percent of this money had actually reached Bhopal, even eighteen months after the accident. At the same time, the government of India claimed that its "sovereign duty" was to procure compensation and provide adequate relief to victims, but in reality was unable to meet victims' most basic needs. Finally, private U.S. lawyers representing the victims vowed that they would see to it that justice prevailed, that Union Carbide was punished for its negligence, and that the Bhopal lawsuits taught a lesson to all multinational corporations. Their actions, however, betrayed a cynical and shameless eagerness to settle the case with Union Carbide's first offer.

In these legal wranglings, victims became hostage to the economic objectives of their own private lawyers and those of Union Carbide, and to the political objectives of their own government. Victims were pawns in complex legal battles being waged supposedly on their behalf.

Historically, the cost of compensating victims in industrial accidents has been very low. It is, for example, only a small percentage of the cost of pollution control.[1] There is no economic reason for making crisis victims wait for the legal system to prove liability in order to be compensated. There are methods both inside and outside court to resolve the compensation conflict.

Even in the Bhopal crisis, for example, interim solutions are possible, so that victims may be compensated while liability is still being determined. By making an out-of-court tentative settlement with victims' lawyers, Union Carbide has acknowledged a liability of *at least* $350 million. By unofficial accounts, the government of India expects a larger amount. While negotiations continue, the judge presiding over the case could order Carbide to pay a certain sum per year (deductible from the final settlement) to finance interim relief and compensation. This order would build on Judge Keenan's earlier precedent of requiring Carbide to contribute $5 million toward interim relief. Even if negotiations took several years, Union Carbide's payments would not exceed the liability it has already acknowledged.

Use of these funds would vary from year to year, but in the initial years they could be used for:

- generating annuities for compensation to survivors of the dead and seriously disabled victims. There are 15,000 to 20,000 such people in Bhopal. They could receive monthly payments that would help them meet their living expenses.

- statistical studies of the victims and preparations for implementing the final settlement. These studies should be done by neutral and independent agencies within the purview of the court where the case is adjudicated.

- alternative employment schemes, nutrition programs, new medical treatment programs, and other services.

Such an interim solution would benefit all parties. The benefits to the victims are obvious. For the government of India and Union Carbide, this solution offers the chance to act ethically and humanely and, in the process, restore their slowly eroding legitimacy. Financially, this solution may help Union Carbide by easing its cash-flow problem. The staggered annual payments will be easier to accommodate than a large, lump-sum payment five years from now. The interim solution might also facilitate a final settlement by bringing Carbide and the government closer together in their respective understandings of the crisis. In short, it would be economically, politically, and morally right.

There is no reason why standing compensation funds cannot be established around the world to offer assistance to victims immediately after a triggering crisis event. Such compensation funds could be financed by corporations in much the same way that insurance systems are funded. Alternatively, governments could create such funds, modeled on the relief funds many countries have to assist in the aftermath of natural disasters such as floods, famines, earthquakes, droughts, and hurricanes. For example, in India, the Prime Minister's Relief Fund, a standing fund designed to assist natural disaster relief efforts, was used to aid the Bhopal victims.

While conflicts are being adjudicated in courts of law, it is the moral duty of all concerned stakeholders to explore nonjudicial approaches to resolving their conflicts expeditiously. Even twenty months after the accident, the judicial process had not even fully resolved the question of where the case should be tried. A complete judicial resolution of the compensation issue could easily take five to seven years. This delay could be avoided by resorting to alternative dispute resolution mechanisms.

The goal for alternative dispute resolution systems is not just to sort out liability issues but also to build an understanding among the parties involved. Court proceedings are adversarial in nature. As such, they should simply serve as a base line for moving the resolution forward. It is important to break down conflicts into constituent parts. Each subconflict must be described in terms of contestants, their respective stakes, and the relative importance of speedy resolution. A victims compensation fund, for example, has the advantage of separating compensation of the victims, which should occur quickly, from the question of legal liability, which may be resolved later.

With the victims' immediate needs taken care of in this way, conflicts between corporations, or between corporations and governments, can be resolved over a longer period of time. But even these conflicts need not be dealt with in the adversarial courtroom atmosphere. Alternative methods have been developed that not only resolve the conflict, but also force the participants to look beyond their own frames of reference. Such methods also have the advantage of being able to accommodate more than two parties.

In the United States, for example, a growing number of environmental disputes are dealt with through mediation. Typically, a third-party arbitrator, mediator, or facilitator tries to sort out the dispute in a nonadversarial, nonlitigious atmosphere. This method has been applied in less than 5 percent of all environmental disputes, and usually succeeds only under certain conditions. The parties must be approximately equal in power and must have a desire to communicate, reach a speedy conclusion, and exchange information openly.[2]

The mediation method might not succeed in the Bhopal case, because of the vast differences in the respective power of the parties involved and because the parties do not trust each other enough to communicate and exchange information openly. But there are other possibilities. One would be to establish a nonpartisan commission consisting of a few people of impeccable integrity. For example, the Nestle infant formula controversy was resolved by a commission headed by Senator Edmund Muskie. (Bertrand Russell headed another like commission dealing with the war crimes of World War II.)[3] A Bhopal commission could include world leaders and humanists who would help resolve disputes concurrently with the judicial process. Such a commission would work using a nonconfrontational, problem-solving approach aimed at resolving disputes within a reasonable length of time.

Methods such as those outlined above can be used as models for resolving industrial crises and for building up the trust required to expand each party's frame of reference.

Actions to Prevent and Cope with Industrial Crises

But it is not enough for the different parties merely to broaden their frames of reference. The parties must also be prepared to take concrete actions and commit resources. Once the various stakeholders gain an understanding of how their individual actions can cause an industrial crisis, they must then take steps to lessen the risk of such a crisis occurring.

Each stakeholder has a different role to play in lessening the possibility of crisis, but, in general, all the stakeholders would do well to follow several general steps. Following these steps requires a willingness to understand the risks of industrial facilities, accept those risks, and deal with them accordingly.

First, *acknowledge the possibility of crisis potential.* Corporations and governments, particularly, are sometimes reluctant to admit that their industrial activities could lead to serious crises. But this sort of ostrich attitude only makes the crisis worse if it *does* occur, because it catches them unprepared.

Second, *freely share information on the potential for crisis.* Usually, a great deal of information on the crisis potential of industrial products and processes is in the hands of the government, corporations, or any number of national or international agencies. Better dissemination of this information to all parties before a crisis occurs would not necessarily incite public hysteria; people are often willing to accept risks, if they know exactly what those risks are and if they believe risks are distributed fairly.

Third, *conduct better "what-if" planning.* Emergency actions following an industrial accident can be far more effective and coordinated if they are planned in advance. Indeed, it is imperative that the corporation, government, and community all work together to plan for possible crises.

Individual stakeholders can also take specific steps to prevent and cope with industrial crises more effectively.

Government

As the protector of public interests, the government has primary and diverse responsibilities in dealing with industrial crises. As the orchestrator of national policies and resources, the government also possesses the power to implement solutions. It can take preventive actions by adopting economic and industrial policies that explicitly attempt to overcome the negative effects of industrial activities, and by ensuring that these policies are implemented scrupulously.

Preventive Actions

The key to the government's crisis prevention effort must be a *sustainable industrialization strategy* that takes into account the hazards of technologies and the ability of local industrial infrastructures to support these technologies.

Technologies must foster *sustainable* economic development. This means that industrial technologies must be suited to the human capital, natural resources, and infrastructural capabilities of the areas in which they are located. Without such economic development, production will cease no matter what technology is used.

The likelihood of major industrial accidents, particularly in developing countries, is almost wholly dependent on national economic development strategies. These strategies determine (1) whether the nation will pursue technologies with limited crisis potential and (2) whether industrial investment will be accompanied by an adequate investment in industrial infrastructure.

Yet, the possibility of an industrial crisis is almost totally disregarded when nations select certain technologies as part of their industrial strategy. Industrial strategies are normally driven by considerations of national resource endowments, market opportunities and threats, and the availability of technology, with virtually total disregard for the technology is negative effects or crisis potential.[4]

Hazard/crisis potential can be used as a screening criterion by incorporating such considerations into industrial licensing procedures at the national and state levels. Technologies with high crisis potential could be denied license or be subjected to closer scrutiny. Similarly, environmental and social impact assessment could be made a part of the licensing process. This would provide a mechanism for assessing the impact of technologies before they are introduced into a community.

The strategic error lies not in the use of a hazardous technology per se but, rather, in the commercial use of such technologies before their crisis potential has been fully examined and methods developed with which to deal with that crisis potential. Government policies should provide a mechanism for restraining private firms until the technology and its supporting infrastructure are sufficiently developed to avoid crises.

While the idea of sustainable development applies to all countries, it has more immediate application in developing countries that are rapidly industrializing. Developing countries have ample supplies of labor and rampant unemployment problems; they have acute infrastructure shortcomings and ubiquitous pollution problems; and they have limited resources for procuring or developing new and appropriate technologies. Yet many developing nations have chosen indiscriminate, urban-based industrialization that simply bypasses 70 percent of the population living in rural areas. The sustained economic health of these countries requires a shift away from this approach to an economic development program that will use available human resources and exploit natural resources less rapaciously.

Sustainable development does not mean deindustrialization or regression to preindustrial modes of production. Instead, it simply requires a creative search for safer, environmentally less stressful, labor-intensive "appropriate" technologies that can be operated on a decentralized scale to fuel more balanced economic and social development. Such technologies are available and, in many sectors, are far more advanced than the technologies now being used. For example, in the energy sector, traditional, centralized sources of power—thermal, hydroelectric, nuclear—can be supplemented with sources capable of producing power on a smaller scale. Indeed, in India, bio-gas, which is produced by each household from cow and buffalo manure, has become a vital source of energy.

It is inevitable that even the most carefully selected portfolio of safe technologies, chosen with the aid of a well-designed, sustainable development strategy, will include some hazardous facilities, such as nuclear plants, chemical plants, toxic-waste sites, and missile bases. These facilities will be chosen despite their hazards because they bring desirable products, jobs, taxes, or a higher standard of living to the area.

Because of this inevitability, *hazard management policies* are needed to ensure that the public is adequately protected.[5] The Seveso Directive, adopted by the Council of the Organization for Economic Cooperation and Development (OECD), an organization of Western European nations, is an example of a governmental policy framework established to deal with major technological hazards.

The Seveso Directive deals with this problem by establishing a two-way interaction between government and manufacturers, while at the same time acknowledging the public's right to know.[6] Although the Seveso Directive assigns only a limited role to the public, it represents a clear improvement over the fragmented, ad hoc regulatory responses to individual crises that we have witnessed in the past.

The World Bank has also begun to use hazard potential as a criterion in evaluating the suitability of projects in developing countries. Development projects funded by the bank are now subject to environmental and safety assessments at all stages of project management. The World Bank also requires corporations and governments to deal with the risks of major hazard installations in developing countries in a systematic way.[7]

Just as important as hazard management policies is a set of *hazard location policies.* Zoning laws frequently create buffer areas between industrial facilities and residential neighborhoods. But these buffer areas are often inadequate because they are not designed for worst-case scenarios. Corporations resist using worst-case scenarios because they are expensive. But as the recent Soviet nuclear disaster suggests, industrial accidents can sometimes affect people who live hundreds of miles away.[8] Therefore worst-case scenarios must be given serious consideration.

Another way of approaching the question of location is to actively involve all stakeholders, not just corporations or government, in the siting of hazardous facilities. The state of New Jersey's Major Hazardous Waste Siting Act,[9] passed in 1981, offers one model for tackling the problem. The law established a siting commission that included representatives of industry, government, environmental groups, and communities. Working together through extensive public hearings, the commission developed broad criteria for siting hazardous waste facilities. The criteria included many considerations traditionally ignored in corporate citing decisions, including (1) providing buffer zones; (2) ensuring the structural stability of the facility; and (3) protecting the environment, in particular groundwater, surface water, wetlands, and air quality.

Similar criteria need to be developed for siting chemical plants, nuclear plants, and other hazardous facilities. If such plans are developed jointly by government, corporations, and communities, each group will have a vested interest in seeing that the plans are accepted and implemented.

The final element in creating a sustainable industrialization strategy is *sound infrastructure development*. As the Bhopal tragedy illustrates, the strength of the industrial infrastructure is perhaps the single most important element in determining the incidence of industrial accidents and the capacity of all stakeholders to deal effectively with such accidents.

In developing countries, where such infrastructure is lacking, there are several alternatives. The most obvious one is to increase the resources devoted to infrastructure. With the economic and political problems involved in reallocating or increasing government revenue, this may be impossible. The best approach in such a case might be to mobilize corporations to develop those parts of the infrastructure most needed for safe industrialization.[10]

A second alternative is to limit the pace of industrialization to levels supportable by the existing infrastructure. This alternative may not be as onerous as it seems if it is coupled with decentralized industrialization policies. China has so decentralized its steel industry that it resembles a cottage industry, with steel manufactured in small collectives. Many developing nations, like China, that have a surplus of labor could pursue such a policy.

A third possibility would be to concentrate industrial development in industrial parks and provide a robust infrastructure to these restricted areas. Although such a solution would be economically efficient it might not be politically feasible because it would widen the gap between industrialized and nonindustrialized sections of the country.

Most countries also have an urgent need to improve their *social infrastructure* because it determines the public's ability to judge technological risks. Government can play a constructive role by passing laws that allow communities and workers in and around industrial facilities to learn more about hazards in their vicinity. In New Jersey, where industrial hazards have become a major public issue, state right-to-know laws allow workers and community groups access to such information as the nature and amount of hazardous material contained within a plant, the toxicity of the material, and the by-products it produces.

Of course, the effectiveness of all these actions depends in large part on *policy implementation,* and failures in this area are endemic to modern governments.[11] In Bhopal, for example, the Indian Labor Department investigated a 1981 accident at the Union Carbide plant that killed one person. Consequent recommendations to improve safety at the plant were never implemented. Many of these recommendations might have mitigated the 1984 disaster.

The issue of policy implementation is too complex to be addressed adequately here. But, briefly, good policy implementation requires resources, information, and political will. In environments with resource shortages, lack of information, and disruptive political pressures, only incremental changes are possible. Incremental improvements can be attained by experimenting with pilot projects, establishing accountability, evaluating performance, and providing monetary incentives to the public and the implementers. Community development programs in India have used pilot projects as a method for testing policies and introducing them slowly.[12] Monetary and material incentives to the public and to health workers have been used to implement family planning policies in India.

Community groups can play a vital role by building policy consensus and by monitoring their implementation. The Chinese "mass line" approach, which involves abstracting particular aspirations of the masses into an overall plan of action, is an example. Leaders and government officials are responsible for gathering the ideas of the people, merging them into an action consensus, and then discussing them with the people so that they can adopt the action consensus as their own. Finally, blatant disregard for policies can be controlled by frequent exemplary punishment of offenders.

Coping Actions

Even if all the preventive actions outlined above were effectively implemented, industrial accidents probably would not be totally eliminated. Thus, we also need to enhance our capacity to cope with them. As the aftermath of the Bhopal tragedy revealed, for the Indian government, coping meant organizing massive relief efforts. Such relief efforts are critically dependent on the capability of two systems: the *crisis information system* and the *medical emergency system*.

In the midst of a crisis, immediate information is needed for rescue, technical damage control, and medical treatment. Additional information is needed for supporting decisions about longer-term rehabilitation of affected persons, legal matters, and safety measures.

Much of this information already exists in various data bases around the world. For example, business associations in the chemical, insurance, nuclear, and waste management industries possess data relevant to hazards in their respective industries. The chemical industry, the health care establishment, and government environmental agencies have emergency response procedures and even provide crisis assistance by telephone. U.S. research organizations, such as the National Toxicology Program, the Center for Disease Control, and the National Institutes of Health, and International groups, such as the World Health Organizations, maintain data bases that could be valuable in an industrial crisis. An effective information network could be created by linking these data bases together and creating additional ones to fill gaps.[13]

Using this information during a crisis presents an even more complex problem. Decisionmakers in remote crisis locations may be unable to gain access to the information if they lack communication lines, electricity, or technical personnel who can operate the system and interpret the information. One solution would be to establish regional and national crisis information teams, consisting of data processing, communications, and language experts who could move quickly into crisis areas with the equipment and skills necessary to assist local authorities, corporate decisionmakers, and the media.

A second problem in regard to information is that some actors in a crisis will try to control the availability of information. For example, the government of India did not release the data it acquired from inside the Bhopal plant, nor did it release the results of its medical research on victims. Such realities merely underscore the importance of creating a crisis team that could serve as a neutral and independent source of information.

Medical emergency systems are similarly strained under crisis conditions, and their requirements are very different in industrialized and developing countries. In industrialized countries, health-care systems are well developed and merely need to be coordinated in order to provide the extra capacity needed to cope with a crisis. The National Medical Emergency System in the United States, designed to handle military casualties in case of a limited nuclear war, is an example of such a system. It involves a network of health-care facilities, each committed to providing a certain number of hospital beds (totalling 500,000) on twenty-four hours' notice. These beds and associated medical services would be mobilized in nuclear emergencies exclusively for treating war-injured patients.

In developing countries, the existing medical and public health facilities are inadequate even for daily life. Here, the development of emergency capacity must begin by strengthening the everyday medical infrastructure. In addition, during crises, resources may be mobilized from international health and charitable organizations. The human costs of *not* developing the necessary emergency capacity were tragically apparent in Bhopal. Medical needs so completely overwhelmed the medical establishment that despite the heroic efforts of individual medical workers, the system was barely able to provide first aid and symptomatic treatment to victims.

Corporations

Corporate plans to prevent crises should go beyond considerations of economic costs and benefits to include a sense of institutionalized professional, ethical, and social responsibility that is acceptable to key stakeholders. Whether it is through a voluntary code of ethics, the use of corporate culture, or the development of personal leadership, corporate strategies need to be reoriented to include a long-term concern with industrial safety, and business unit strategies need to deal with operational improvements that will make products and facilities safer.

Preventive Actions

Corporate strategies will not become more sensitive to crisis prevention until corporations begin to develop socially responsible corporate cultures or frames of reference.

Corporate culture has been defined variously as shared values and assumptions; learned behaviors; collective mental programs; and corporate ideology or belief systems. Corporate cultures embody the key objectives and values that guide corporate decisions. Preventing crises requires the development of socially responsible corporate cultures that are as sensitive to safety and environmental concerns as they are to economic and technical matters.

For our purposes, a more useful method of understanding corporate culture is to view it as a corporation's frame of reference. Since frames of reference include basic values and assumptions, they include a company's culture. Thus, they provide a means of understanding the influence of corporate culture on decisions.

As stated in the previous chapter, a corporation's frame of reference is used to legitimize a narrow strategic thrust concerned with profit and efficiency. Operating and safety decisions are made within the context of this narrow thrust. Some of these individual decisions are not optimal, while others have negative effects on safety or the environment. Managers are willing to overlook these negative effects in order to conform with the strategic thrust and with the corporate frame of reference.

Thus, the key to safer business practices is creating a socially responsible corporate frame of reference that is highly sensitive to safety, environmental protection, and the hazard potential of operations. Corporate managers could take the following steps to reshape their corporation's frame of reference:

Systematically question assumptions underlying organizational policies about safety and environmental issues. If formal policies do not exist, examine specific decisions that deal with these issues in order to identify assumptions. Evaluate these assumptions based on their factual and economic validity and on their moral and social acceptability.

Modify decisionmaking processes on safety and environmental issues to include the concerns of governmental agencies, workers, and communities. A dialogue group including these stakeholders can be set up to act as a vehicle for identifying these concerns.

Make safety (for both workers and the community) and environmental protection a strategic issue for the organization. Develop specific strategic objectives, programs, and evaluation/compensation systems in these areas. Make top management responsible for achieving these objectives and provide the resources necessary to achieve them.

Reassess the cognitive maps and the causal reasoning that guide the corporation's methods of identifying problems and finding solutions with regard to industrial and consumer safety, occupational health, and environmental protection. Expand the causal analysis of these issues beyond mere technological variables to include social and ethical issues as well.

The creation of a socially responsible corporate frame of reference, or culture, is more difficult in multinational corporations that operate in many cultural environments and employ people with vastly different backgrounds. But structured corporate-wide approaches are possible. Corporations could, for example, adopt international codes of conduct with respect to health and environmental issues.[14] These codes are written in very general language and only a few of them address environmental issues.[15] But these codes do attempt to equalize competitive differences of domestic and foreign investors, guide corporate activity in socially desirable directions, and minimize disputes between corporations and their host countries.

Even if a corporation does not adopt the international codes, it can adopt consistent policies—rather than the traditional "double standard" approach—on safety and environmental issues. Dow Chemical, for example, has written minimum requirements on safety, loss prevention, worker health, and environmental matters for its 425 processing plants around the world. Local standards may be more rigid than these minimum requirements, but they cannot be more lax.

A socially responsible corporate frame of reference can also be built by giving greater attention to human resources. Human mistakes are often at the heart of industrial disasters. While human error will surely never be eliminated, it can be greatly reduced through better technical and safety training, as well as "what-if" training that anticipates system failures.

However, building a more socially responsible corporate culture is not enough. Corporations must take many other preventive steps. Perhaps the most important of these is the development of *safe technology portfolios.* Like governments, corporations typically make technology choices based on economic and market considerations. In the case of multinational corporations attempting to diversify geographically, these considerations include the international product life cycle, foreign market needs, technology transfer arrangements, host country regulations, and international competition. Safety is a secondary, technical consideration, rather than a primary consideration in strategic decisions.

Corporate technology portfolios can be made safer by screening technologies according to a set of criteria that assess their crisis potential. Conceptually, this can be done by using the framework shown in Figure 6-1. The first step is to perform an environmental impact assessment or worst-case scenario analysis in order to rate the crisis potential of each new hazardous facility. In addition, an infrastructure-capability audit should be conducted to rate both the physical and social infrastructure. A list of additional infrastructure needs can be conveyed to local authorities, though some, as we will discuss later, may be developed by the corporation itself.

For most corporations, the first step toward institutionalizing safety in and around industrial plants is *environmental impact assessment* (EIA) of both new and existing facilities. An EIA would force managers to examine explicitly the adverse effects of a facility and identify problems that could trigger a crisis. It would also generate information that can be used in many ways—for safer construction and operation, for procuring necessary government permits, for monitoring its hazard potential, for aiding in technology transfer, and for resolving conflicts in case of disputes.[16]

All too often, of course, EIAs are undertaken as a pro forma exercise designed to fulfill some procedural requirement.[17] To be more effective, they could be systematically tied to annual planning and budgeting cycles. The resources required to act on EIA conclusions must be made available. If EIAs are done correctly, they can identify problems and reduce the chance of an accident.

One area the EIA might identify is needed *infrastructure* facilities and services. Corporations can reduce the probability of an accident by assisting governments in developing facilities. This is

Figure 6–1. Technology Portfolio Screen.

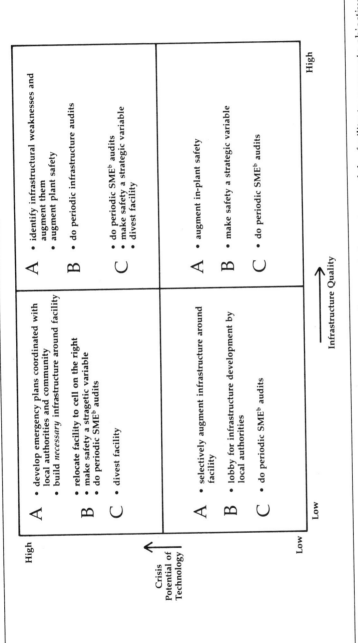

a. Facilities are represented by A, B, and C. A represents high strategic importance of the facility to corporate objectives, B moderate, and C low. General suggestions for dealing with different types of facilities are given in each cell.
b. Safety, maintenance, and environmental.

particularly true in developing countries, where the government either does not possess enough resources or technology, or will not build the infrastructure because of its own economic and political priorities. Corporations should seriously examine ways to develop or promote the development of needed infrastructure.

There is a long history of corporate involvement in infrastructure development. In the past, entire townships have been constructed by corporations in conjunction with local authorities. Company towns in the United States and housing colonies owned by public and private corporations in India are good examples of this precedent.

This does not mean that corporations should pay for basic infrastructure for the entire town. Rather, it implies that they should selectively develop the services that are needed for safe operation of their own industrial facilities and that primarily benefit their own workers and business associates.

For example, companies can erect fencing around hazardous facilities in order to prevent unauthorized persons from getting close to them. They can build sewage extensions to connect their plant with the town's sewage system. They can provide safe housing for their own employees, near the plant but with sufficient buffer zones. They can buy the land around the plant and leave it vacant to prevent people from living in dangerous proximity to it. The specifics, of course, are contingent on particular circumstances. But the responsibility lies with corporations to operate their plants in a safe environment.

Another area in which the EIA can assist is in helping corporations to broaden the scope of their *hazardous siting decisions*. Businesses traditionally use economic and technical criteria for locating facilities. These criteria include the cost of transporting raw material to production facilities and finished goods to markets, the local labor market, the availability and price of industrial infrastructure, the quality of life for employees, and state and local tax incentives. Siting decisions are made at the project design stage by corporate engineers and managers.

But hazardous facility location is as much a political decision as it is an economic and technical one. Siting decisions for such facilities must involve the stakeholders likely to be affected by crises at the facility. At the very least, government and community representatives should be consulted on these decisions. Ideally, all three parties should work together, as they do under the terms of the New Jersey law mentioned above.

Like EIAs, *safety audits* can be extremely important, but they too are often done simply to fulfill procedural requirements. To avoid this problem, top management must pay attention to safety and provide resources to support it. This approach can be institutionalized by establishing a top management accountability and review system separate from the audit department (which is, in turn, separate from the facility's operating management). The reward system for operating management should be linked to the safety performance of the plant.

A second problem with safety audits is that most often there is no dynamic updating of safety standards and risk assessment, which, in turn, makes safety audit procedures obsolete. New safety technologies might be available and the risks posed by the plant may change with time. Bhopal was a classic example of this problem. First, new computerized safety systems were available after the plant opened, but they were not incorporated at the plant. Second, the increased risk posed by the establishment of an MIC production unit at the plant in 1980, and the concurrent establishment of slum colonies around the plant, was never recognized. Dynamic updating of such changes should be built into organizational procedures for safety audits. Risk assessments should be done periodically, and especially after major changes in or around the plant have occurred.

In addition to safety audits, *environmental audits* that analyze compliance with regulatory standards are necessary for legal protection. In the absence of regulatory standards, voluntary corporate standards may be used to assess environmental damage and crisis potential. However these assessments are made, they should form the basis for remedial action to make the plant safer or to relocate the plant in a safer location.

In addition, most industrial plants can make *operational safety improvements* that are relatively inexpensive but that significantly reduce the risk of accidents. For example, more information on hazards can be provided to relevant community and government authorities so they are better equipped to prepare for emergencies. Also, safety alert procedures can be changed to assure that the community is warned of a potential emergency as soon as it is suspected.

Several routine decisions related to production scheduling, inventory management, materials control, and operating procedures increase the likelihood of a crisis erupting because they are based on economic and technical efficiency, rather than plant safety. Decision criteria must go beyond technical and economic variables to include social and ethical variables as well.[18] Like many of the other changes discussed here, this broadening of criteria must be built into all corporate policies and procedures.

Corporations can also use two methods to *enhance the operating skills of personnel* and, thus, reduce the chance of accidents. First, they can rotate industrial plant personnel through a variety of jobs, thus giving workers an understanding of the entire system and its internal interconnectness. This might have reduced the possibility of an accident in the Bhopal plant, where the operators did not understand trouble indicators because they did not understand the interconnectedness of the system. Rotation might also keep employees from being desensitized to dangers and treating hazards in a routine fashion, a habit which in itself poses a considerable threat.

Second, corporations should provide ample human backup for operating systems that have crisis potential. Redundancy in parts and subsystems is a well-established principle in designing manufacturing facilities. Redundancy in personnel, on the other hand, is considered wasteful. Yet in some circumstances, personnel redundancy may enhance plant safety. Some plants, such as the Union Carbide plant in Bhopal, are designed for manual operation; others rely heavily on human messengers for communication. Such plants are relatively common in developing countries. In these situations, safety is critically dependent on continuous human intervention. Extra personnel in these locations act as safety buffers in case of emergencies.

Coping Actions

Because corporations do not control all the parameters that create crisis conditions, they will never be fully protected from crisis. Thus, planning to cope with the inevitable crisis is an urgent corporate need. Corporate crisis management can be streamlined in many ways to ensure thoughtful responses to crisis situations.

The lack of "what-if" *emergency planning* is often a major barrier to coping with crises effectively. The likelihood of a major crisis is so low that it seems unreal to most people. Nevertheless, well-designed and well-maintained emergency plans can prevent accidents from causing great damage. The ability to view the unreal as possible—and the possible as probable—requires a level of intuition and imagination that many corporate managers lack.

Plant-level emergency planning depends on local conditions, but it can be enhanced in several ways. Normal emergency procedures usually focus on likely emergencies, on protecting workers in the plant, and on protecting personnel in the immediate vicinity. These procedures may be augmented with procedures for worst-case scenarios or low-probability catastrophic events, so that community concerns and secondary effects are taken into account.[19] Communities and local authorities should participate in developing emergency plans, so they will understand potential hazards. Community agencies should be encouraged to develop their own off-site emergency plans.

One good example of community-based emergency planning is Louisiana's planning for the transportation of hazardous materials, which is coordinated by the state police. Louisiana is among the largest producers of hazardous materials in the country, and it is estimated that 25 percent of all hazardous material produced in the United States is transported through Louisiana. The state's emergency response program involves community groups, government, and private businesses. It provides a variety of technical assistance, training, equipment, and necessary information to community members.

Industrial corporations are often reluctant to release hazard-related information. They claim that doing so compromises the secrecy of competitive data and unnecessarily alarms the community. Both excuses are untenable. Typically, the information needed for hazard control and emergency planning by communities is too fragmented to be of competitive value. It concerns only limited aspects of hazardous production materials and facilities. In regard to community reaction, alarm is not caused by the hazard potential *per se* but, rather, by the realization that there may be no plans to cope with it. This issue can be handled by adopting emergency procedures that would communicate risk information to the public *and* assist affected areas in coping with a potential crisis.

Corporations would also benefit from the development of *crisis coping teams and processes*. Crisis conditions often paralyze otherwise robust management teams, leading them to defensive and reactive responses. In Bhopal, for example, UCIL stonewalled the press, and the parent company tried to distance itself from its subsidiary's management—even though both responses harmed the crisis management effort.

The initial response to the crisis sets the tone for the rest of the effort. If it is not handled sensitively, multiple conflicts erupt rapidly and aggravate the situation, as was the case in Bhopal. In addition to technical damage control, top priority should be given to rescue, relief, and other *human* and *social* aspects of the crisis. Solving the human problems wins public support and allays the suspicions of stakeholders. The worst mistake corporations make in the initial phase of a crisis is to withdraw from other stakeholders and act defensively—sometimes even denying that something is wrong. Such defensive activity is misrepresented and misunderstood, and it generates suspicion. It makes rescue and relief operations difficult and ineffective.

To overcome these tendencies, crisis coping teams are needed. These should include specially trained managers, communications experts, and people familiar with the technical and social environment of the crisis site. Crisis coping teams should be actively involved in identifying crisis issues and solving the ad hoc problems that vary from crisis to crisis.[20] The success of solutions depends as much on their acceptance by important stakeholders as on the technical efficiency with which they are carried out. Thus, it is critical to establish a network of representatives from the various stakeholder groups to serve as a forum for testing crisis response ideas.

Later in the life cycle of a crisis, the focus shifts from crisis management to conflict resolution. As discussed earlier, the use of alternative methods for dispute resolution should be explored. This may require mobilizing the good will and mediating services of uninvolved, independent, and credible third parties. For example, the Nestle Infant Formula Audit Commission, headed by Senator Edmund Muskie, was able to solve the lengthy dispute over the use of infant formula in developing countries by mediating between Nestle and other infant formula manufacturers and activist groups who had organized a worldwide boycott of infant formula products.

Johnson & Johnson's handling of the Tylenol poisoning crisis in 1984 illustrates some elements of an effective crisis coping team. The crisis team was headed by the company's chief executive officer, James Burke. Instead of distancing itself from its manufacturing subsidiary, McNeil Laboratories, and thus shifting the blame, the Johnson & Johnson team internalized the subsidiary's problems. The company enacted a massive recall campaign; began an extensive public relations effort, sensitive to consumer-safety concerns; and maintained continuous and open communications with confused dealers.

As the Johnson & Johnson experience illustrates, *public relations and communications management* is of great importance because crisis events invariably become the center of media attention. Public relations efforts must be credible. They should be aimed not just at presenting the corporation in a favorable light, but at clarifying issues and at presenting the corporate position on central conflicts.

Public relations campaigns require delicacy and sensitivity, because they must simultaneously reach many different audiences, including stockholders and financial analysts, governmental agencies, judges, foreign governments, victims and those assisting them, and the general public. In addition, the corporation must make a concerted effort to provide information to its employees, particularly those at the crisis site, and to business associates who have transactions in process at the time of the crisis.[21]

The most powerful vehicle of communication—and the one most likely to shape general opinion on the crisis—is, of course, the public media. Providing information to the media is a delicate art, frought with opportunities and problems. In the immediate aftermath of a crisis, normal communication channels break down. The press is eager to get the full story, but few facts have been established and the opportunities to verify information are limited. At this point, it is absolutely essential for corporations to provide technical information that is factually correct and also the objective data required by reporters for their stories. Stonewalling information probably will not succeed, because in such situations the public interest stakes are so high that the media will eventually uncover the information. Delaying the release of factual information only antagonizes the press and raises public suspicion.[22]

Communities

Community residents are usually left out of the information loops and the decisionmaking processes concerning industrial crisis prevention and coping actions. Corporations and governments often argue that critical decisions about hazardous industries should be trusted only to technically qualified individuals, suggesting that community residents possess neither the expertise nor the resources to make these decisions.

While such a conclusion may be empirically accurate, risk assessment is not simply the quantification of hazards in probabilistic terms. It also involves judgments and choices about the acceptability of risks and about the equity of distribution of benefits and risks.[23]

Not surprisingly, community residents perceive risks fairly accurately if they are aware of the dangers they are exposed to. Moreover, people are willing to take risks if they believe that risks and benefits are equally distributed across social, economic geographical, and racial boundaries.

In democratic societies, people have the right to choose the risks they live with. Yet risk management is a political process currently dominated by corporations and government agencies. Even in the industrial democracies of the West, risk decisions are most often made by governments and corporations, with little community participation.[24] Even in the United States and Europe, where grassroots environmental movements are strong, public participation is limited by a lack of resources and legitimacy, as well as by the adversarial relationships that develop between environmental groups and other stakeholders.

Of course, there is a moral argument for bringing community residents into the decisionmaking process. While they are the most directly affected group in any industrial crisis, their interests are often only indirectly addressed by other stakeholders. As stated at the beginning of this chapter, community action is often an effective means of changing governmental and corporate crisis procedures. Beyond these more idealistic arguments, however, there are important practical reasons why corporations and governments ought to involve communities into the decisionmaking process.

Centralized risk management procedures—whether established by governments or corporations—usually lack the resources to monitor risks in remote locations. Community residents possess first-hand information from their own experience about environmental conditions and industrial hazards in their vicinity. They are also in touch with any changes in these conditions that might create the need for action. Communities can be instrumental in preventing crises by identifying dangerous situations early enough for government and corporate decisionmakers to take remedial actions. However, community groups cannot play an effective role in preventing and coping with industrial crises, unless they are provided with the information and resources to take action.

Preventive Actions.

Communities can anticipate and prevent industrial crises, but only if they have highly developed *community awareness* about industrial risks. In Bhopal, residents living around the UCIL plant were warned about its hazardous potential by a series of articles in a local newspaper. Residents ignored these warnings because they did not know how to react to them, while local officials dismissed them as sensationalist reporting. They lacked the community awareness that might have motivated them to force the government and Union Carbide to take the necessary preventive steps.

Community awareness can be developed by systematically educating residents about the nature of technological hazards and through a sustained dialogue among plant personnel, local emergency management authorities, and community residents. Planned and structured dissemination of information would avoid alarming residents. Permanent dialogue groups, whose scope and structure would be tailored to local conditions, would serve a useful purpose.

Mere knowledge and discussion is not enough, however. For residents to serve effectively, *communities must be empowered* by the provision of legitimate forums and resources. This requires passage of such legislation as "Right to Know" and "Right to Act" laws, which allow communities to extract and use information needed for meaningful dialogue. But initiative for such legislation will not come from the government and private corporations. It must come from the people.

Coping Actions

Industrial crises create panic, chaos, highly emotional conflict, and psychological trauma, all of which prevent residents from evaluating their options rationally. Crises shake community confidence in the ability of both government and corporations to prevent crises. People are overcome by suspicion and powerlessness, which, in turn, fosters apathy.

Community organizations must address crises from a unique perspective. They cannot be wedded to the actions and solutions of the other stakeholders. They must address those crisis problems that other stakeholders have neglected or addressed inadequately. There is an urgent need to reestablish that sense of community that enables residents to work together.[25] In Bhopal, the victims' aid groups conducted independent investigations, provided specialized medical and relief services, and successfully raised the political consciousness of victims. This provided victims with a much-needed sense of accomplishment.

Perhaps the most basic requirement of active communities is the creation of independent *community leadership* from within, particularly if existing community leaders and politicians are closely aligned with government and corporate interests. Outside organizers and activists can play limited roles in providing information to an affected community, but initiative and leadership must be indigenous if a sustained effort is to succeed.

Community leadership during a crisis has different requirements than leadership needs at other times. In traditional leadership roles, leaders influence others through personal charisma, expertise, formal authority, or control over resources. Leadership is an exercise in exerting power in well-structured situations to achieve objectives.

In crisis-ridden communities, however, residents are not members of a formal organization. They are operating under a great deal of stress and with different assumptions about and knowledge of crisis events. Often they do not know or cannot agree on what problems are salient and how they can be solved. Crises are highly politicized events and many special-interest groups will struggle to shape their outcome.

Under these conditions, leaders must act to *release* the power of community members, rather than attempt to exert power over them. Leadership lies in giving people the opportunity to express themselves and the encouragement to act. Leaders must be capable of eliciting and communicating information to and from people with widely varying backgrounds and educational levels. They must also be able to build small, cohesive work groups targeted at specific tasks and to facilitate work within and between these groups.[26]

In the case of any industrial crisis, the creation of *local, national, and international networks* of concerned people can help build solidarity around crisis issues. The effectiveness of such networks in keeping the community perspective alive has been proven time and time again. At Love Canal, the local homeowners association and an action group initiated networking among citizens. In Minamata, Japan, where hundreds of people died from eating fish polluted with mercury, the Solidarity Network Asia Minamata initiated a citizen response, which conducted medical studies of the disease, disseminated information, and lobbied for relief support. These activities continue to date. Several citizen groups helped initiate citizen involvement in Bhopal.[27]

The purpose of these networks is to articulate and consolidate an independent community/citizen perspective on the crisis. In democratic societies, the public's view is supposedly represented by the government, eliminating the need for an independent citizen's perspective. This may be true in regard to some public issues, but industrial crisis situations are rife with controversy and conflicts in which the government, as a social institution and as a set of organizations, has vested interests that are different from those of the public. Citizen groups can voice victims' views and legitimate their consideration by decisionmakers.

Local citizen networks often are established around issues such as inadequacies in relief, rehabilitation, compensation, medical treatments, and precautionary measures. They have short lives because they lack resources, focus on immediate needs, and are often overwhelmed by government and corporate responses. Over time, they are branded as radical or marginal, losing legitimacy with the public at large.

To sustain themselves, these networks must project the implications of local issues for other communities nationally and internationally. They must broaden debate over local problems to include national and international policy questions. A clear opportunity for such groups is to link with other consumer, environmental, and labor groups in their own countries, and with similar international organizations.[28] The Green Party in West Germany, which has built a national constituency around local issues of citizen awareness, has made this transition well.

These broader citizen networks can provide a forum for communities to share crisis experiences with each other. By sharing, residents can see commonalities, which will allow them to normalize postdisaster disruptions and focus on only those issues that are in need of resolution. An affected community can regain direction and community spirit by discussing and shaping crisis responses.

One of the most important concerns for any community group involved in an industrial crisis is the *just resolution of conflicts,* so the victims will feel that their problems have been handled in an equitable, timely, and adequate manner.[29] This has been one of the most serious ongoing problems in Bhopal, where victims—already traumatized by the accident—now feel victimized by slow resolution of the compensation issue, which they cannot control.

As the Bhopal accident illustrates, litigation is costly and time consuming, and takes control away from the community, giving it, instead, to lawyers. Such an arrangement is often necessary as a means of exerting pressure on corporations and government, but it fails to achieve timely and equitable justice.

To generate governmental action, communities must exert political pressure. Lobbying and publicity efforts, however, must be focused on concrete demands. As for corporations, communities can exert economic pressure on them. Consumer boycotts, for example, affect revenues and profitability.

Boycotts may be initiated around products associated with the crisis, because they will capture the public's imagination. (Often, the need for these products is already being questioned by consumers.) From this initial group of products, the boycott could be expanded to include products that are most profitable to the corporation. To be effective, boycotts also must be expanded to national and international markets, which will require support from worldwide consumer, church, and public interest groups.

One model for an effective strategy was the Nestle infant formula boycott. Use of infant formulas in the developing world had caused widespread death and malnutrition among infants, and Nestle held the largest share of the market. The conflict lasted well over a decade, but ultimately the boycott succeeded. The case illustrates that successful boycotts—though they require international coordination, money, and time to organize—can do enough financial damage to force negotiations.

A second means of influencing corporations is through shareholder pressure. Shareholder resolutions and lawsuits bring problems to the attention of directors and exert pressure on corporate management to be responsive to crisis problems. This tactic is more likely to succeed in those cases where the management is clearly at fault. Such pressure may embarrass management and highlight the company's responsibilities or negligence, but generally it has limited power to elicit action.

Surviving Industrial Crises

As unbelievable as it might seem, several arguments were proposed that actually sought to diminish the importance of the death and destruction created by the Bhopal accident. One argument popular among rationalists was a cost-benefit analysis, which suggested that modern technology (especially pesticides, drugs, and chemicals) has benefited countless more people than it has harmed, and that Bhopal-like disasters are simply part of the price society pays for the blessings of technology. A second, related argument — the technological imperative argument — says that accidents are inevitable because there are no zero-risk technologies, and we should thus seek improved technological solutions.

The most untenable argument used to diminish the human significance of the Bhopal tragedy — and one used by Union Carbide's lawyers — is the low value of life argument. This argument accentuates the fact that large numbers of people in developing countries die from hunger, malnutrition, and easily preventable diseases. Therefore, lives in the developing countries have a low value and are somehow expendable, unimportant, and really not worth getting alarmed about.

In a world where corporations, governments, and individual communities all claim to be striving toward improving the human condition, it is shameful that these arguments were presented at all, particularly the last one. All three, of course, reflect the narrow frames of reference of different stakeholders in the Bhopal crisis.

The cost-benefit argument ignores the fact that costs and benefits are not equitably distributed throughout society. In Bhopal, those who reaped the benefits of the UCIL plant — workers, rich farmers, stockholders, and government agencies — escaped unharmed, while the slum dwellers who lost their lives benefited little from the plant's presence. The technological-imperative argument ignores the undeniable fact that technological crises are often the result of human, organizational, and environmental factors that have little to do with technology. Far more knowledge is available than is used to make technologies safe. But putting knowledge to use is expensive. Economics and politics, not technology, determine industrial safety. The low value of life argument is so hideous that it is difficult to respond to it rationally. It is a value statement about the worth of human life that can be—and should be—countered only by other value statements that give life greater importance. The kind of thinking that underlies these arguments must be eradicated before we can hope to improve our prevention of and coping with industrial accidents.

Throughout this book, we have seen the importance of breaking down old barriers among different stakeholders in a real or potential industrial crisis. Because the stakes are so high, and the consequences so tragic, all stakeholders—corporations, governments, and communities—are reluctant to open their doors. But with each new tragedy, the reasons for cooperating become more compelling. Very simply, stakeholders need each other. Corporations need the relief assistance and information that governments can provide, as well as the "early warning" and cooperation of communities. Governments need the technical advice and (in the many developing countries) the financial assistance of corporations, as well as grassroots help from communities. Communities, the most easily victimized of the three stakeholders, need information, resources, and cooperation from the other two stakeholders to survive industrial crises.

It will not always be a cordial partnership. These three groups have developed an animosity and distrust toward one another that will not die easily. But theirs is a necessary partnership. Bhopal showed us the dark side of technology. It illustrated how industrial progress can be a Faustian bargain. Future Bhopals can be prevented only through the joint actions of corporations, governments, and the public. They must work together and watch over each other's shoulders to ensure that we do not get ahead of ourselves in our race for progress.

Notes

Chapter 1

1. The event has been called an "accident," although that may not be the best descriptor. An *accident* connotes an unintentional chance occurrence arising from unknown causes. The event was not an accident in this sense of the word. A second meaning of *accident* is an unfortunate event resulting from carelessness, unawareness, ignorance, or a combination of the three. It is in this sense that the word *accident* may apply to the events that occurred at the Union Carbide (I) Ltd. plant in Bhopal.

Chapter 2

1. Social system crises are rooted in economic crises caused by the inherent contradictions of capitalist industrial economies. Economic crises give rise to legitimacy crises, which involve the withdrawal of mass loyalty from the state. These crises erode system integration. Persistent disturbances in system integration impair the consensual foundations of normative structures of social systems. This leads to the breakdown of value systems and institutions that maintain system integration—to anomie. To understand crisis it is important to examine the connections between this disruption of normative structure and system control failures. J. Habermas, *Communication and the Evolution of Society*, trans. Thomas McCarthy (Boston: Beacon Press, 1979); C. Offe, *Contradictions of the Welfare State* (Cambridge, Mass.: M.I.T. Press, 1984).
2. J. Marrone, "The Liability Claims Experience of the American Nuclear Pools and Their Response to the Three Mile Island Accident" (Paper delivered at the OECD Symposium on Nuclear Third Party Liability and Insurance, Brussels, September 1984).
3. J. G. Kemeny, *Report of the President's Commission on the Accident at Three Mile Island* (New York: Pergamon Press, 1981); D.L. Sills, C.P. Wolf, and V.B. Shenanski, eds., *Accident at Three Mile Island: The Human Dimension* (Boulder, Colo.: Westview Press, 1982).
4. D.M. Berman, *Death on the Job* (New York: Monthly Review Press, 1978); N.A. Ashford, *Crisis in the Workplace: Occupational Disease and Injury* (Cambridge, Mass.: M.I.T. Press, 1976).

5. D. Nelkin and M.S. Brown, *Workers at Risk* (Chicago: University of Chicago Press, 1984).
6. The construction of the Gauley Bridge hydroelectric tunnel in West Virginia in 1930–32 exemplifies how crisis emerges in the occupational health area. The tunnel was cut through a site where the rock had very high silica content, as evidenced by tests done by the contractors for the project, the Rinehart–Dennis Company. Silica was known to cause silicosis, a disabling and incurable lung disease. Working conditions in tunnel construction sites are usually poor. At the Gauley bridge site, dust was so thick that workers could barely see ten feet under train headlights. Hazards were further exacerbated by a lack of protective equipment for workers and by new, inexperienced workers from southern states who replaced local miners who had quit working under such hazardous conditions. Within nine to eighteen months of exposure to the silica dust, workers began dying. An estimated 476 workers died and 1,500 were injured during the construction project (D.M. Berman, *Death on the Job*). Rinehart–Dennis was sued for negligence under common law by workers and their relatives. One hundred sixty-seven suits were settled out of court for a total of $130,000.

    Another example of chronic harm from industrial production that resulted in crisis for both workers and consumers is asbestos-related injuries. From 1940 to 1980, nearly 21 million Americans, and a far larger number of workers abroad, were exposed to asbestos fibers. Microscopic asbestos fibers entered the lungs and caused cancer in a large proportion of these people, of whom 8,000 to 10,000 will die annually until the year 2000. Thousands of others will suffer related lung diseases (Paul Brodeur, *Outrageous Misconduct: The Asbestos Industry on Trial* [New York: Pantheon, 1985]). Worldwide, asbestos-related deaths cannot be estimated—they can only be imagined. While asbestos is now banned in the United States, it is commonly used in construction in many developing countries. In most of these countries, workers and users possess very little information on its hazardous character, despite the fact that for at least the past fifty years its carcinogenic properties have been known.
7. L. Brown et al., eds., *State of the World 1985*, a Worldwatch Institute Report [New York: W.W. Norton and Co., 1985]) describes crises created by depletion of fresh water supplies. Destruction of forests by acid rain threatened the extinction of global fisheries and of biological genes. Our vulnerability to the energy crisis was evident during 1973–74, when rising oil prices suddenly threw the world economy into a recession.
8. A.G. Levine, *Love Canal: Science, Politics, and People* (Lexington, Mass.: Lexington Books, 1982).
9. M.K. Tolba, *Development without Destruction: Evolving Environmental Perceptions* (Dublin: Tycooly International Publishing, 1982).
10. Center for Science and Environment, *The State of India's Environment: A Citizen's Report* (New Delhi: Center for Science and Environment, 1982).
11. B.I. Castleman, "The Export of Hazardous Factories to Developing Nations," *International Journal of Health Services* 9 (1979): 369–806.
12. M. Simons, "Some Smell Disaster in Brazilian Industry Zone," *New York Times* 18 May 1985, p. 2.

13. S. Kinghorn, "Corporate Harm: An Analysis of Structure and Process" (Paper delivered at the conference on Critical Perspectives in Organizational Analysis, Baruch College, City University of New York, September 5–7, 1985); I.I. Mitroff and R.H. Kilman, *Corporate Tragedies: Product Tampering, Sabotage and Other Disasters* (New York: Praeger Publishers, 1984).

14. R. Norris, ed., *Pills, Pesticides, and Profits* (Croton-on-Hudson, N.Y.: North River Press, 1982).

15. J. Pearson, *Technology, Environment, and Development* (Washington, D.C.: World Resources Institute, 1984).

16. L. McLaughlin, "Concern Is Growing in Peru over Use of Banned Chemicals," World Environment Report, Washington, D.C., October 8, 1979.

17. U.S. General Accounting Office, *Better Regulation of Pesticides Exports and Pesticides Residues in Imported Foods is Essential* (Washington, D.C.: Government Printing Office, 1979); D. Weir and D. Shapiro, *Circle of Poison* (San Francisco: Center for Investigative Reporting, 1981).

18. E. Eckholm and J. Scherr, "Double Standards and the Pesticides Trade," *New Scientist* 77 (1978): 441–43; R. Norris, ed., *Pills, Pesticides and Profits* (Croton-on-Hudson, N.Y.: North River Press, 1982).

19. R.D. Pagan, "The Nestle Boycott: Implications for Strategic Planning," *Journal of Business Strategy* (Spring 1986): 12–19; E.S. Muskie and D.J. Greenwald, "An Overview of the Nestle Infant Formula Audit Commission: Is It a Model?" *Journal of Business Strategy* (Spring 1986): 19–23; S.P. Sethi, Hamid Etemad, and K.A.N. Luther, "New Socio Political Forces: The Globalization of Conflict," *Journal of Business Strategy* (Spring 1986): 24–31.

20. C.E. Fritz, "Disasters," in *Social Problems*, ed. R. Merton and R. Nisbet (New York: Harcourt Brace and World, 1961), pp. 651–94; E.L. Quarantelli, ed., *Disasters: Theory and Research* (London: Sage, 1978).

21. K.T. Erickson, *Everything in Its Path* (New York: Simon and Schuster, 1976).

22. G.A. Kreps, "Sociological Inquiry and Disaster Research, *Annual Review of Sociology* 10 (1984): 309–30.

23. B.A. Turner, "The Organizational and Interorganizational Development of Disasters," *Administrative Science Quarterly* 21 (1976): 378–97.

24. H. Smets, "Compensation for Exceptional Environmental Damage Caused by Industrial Activities" (Paper delivered at the conference on Transportation, Storage and Disposal of Hazardous Materials, IIASA, Laxenburg, Austria, July 1–5, 1985).

25. B.A. Turner, *Man Made Disasters* (London: Wykeham Publications, 1978).

26. K. Weick, *The Social Psychology of Organizing* (Reading, Mass.: Addison-Wesley, 1979).

27. P. Shrivastava, "Organizational Myths in Industrial Crises" (Working Paper, Industrial Crisis Institute, New York, N.Y., May 1985).

28. A. Giddens, *Central Problems in Social Theory* (Los Angeles: University of California Press, 1979); J. Habermas, *Communication and the Evolution of Society*; C. Offe, *Contradictions of the Welfare State*.

29. Various branches of the state are compelled simultaneously to perform two incompatible functions in managing the economic system: commodification and decommodification. *Commodification* requires the state to give private capital and the power and autonomy to invest in and organize economic pro-

duction, and to accumulate capital. Welfare state administrators have a self-interest in giving preferential treatment to the capitalist economy and preserving the private sector's power and the scope of commodification processes to avoid economic crisis and thereby retain mass loyalty. However, commodification needs to be controlled by the state through processes of *decommodification* (nonmarket intervention). The state is thus put into the contradictory position of being self-limiting and subordinating itself to the economy, while occasionally intervening through nonmarket (decommodified) means to create preconditions for its existence and successful functioning. Thus, the welfare state acts as the "obvious-invisible hand" in relation to the economy (Offe, *Contradictions of the Welfare State*).

30. J. Habermas, *Theory of Communicative Action*, 2 vols. (Cambridge, Mass.: M.I.T. Press, 1983).
31. Castleman, "The Export of Hazardous Factories."
32. World Bank, *World Development Report* (New York: Oxford University Press, 1984).
33. The economies of most developing countries have a dual technological and economic character. They have a modern sector that features large-scale, capital-intensive industries using modern product and process technologies and employing a very small percentage of the total labor force. The larger part of the economy is still in the traditional sector that is primarily agricultural, with small-scale, labor-intensive, and simple and primitive technologies.
34. M. Lipton, *The Urban Bias* (London: Longmans, 1976).
35. In the second half of the 1970s, several official estimates of the toxic waste clean-up costs were made. The Environmental Protection Agency (EPA) estimated that over 10,000 toxic waste dumps in the United States need cleaning up. It compiled a list of 825 sites deemed to pose long-term threats to human health and the environment. This National Priority List is expected to grow to over 2,500 in the next few years. The EPA estimated the cost of clean-up to be about $23 billion spread over eight to ten years. These estimates have been criticized as grossly underestimating the problem. The Office of Technology Assessment (OTA), for example, estimated that clean-up is needed at nearly 10,000 uncontrolled hazardous waste sites. This could take up to fifty years and cost more than $100 billion. And the General Accounting Office (GAO) projects that the National Priority List could eventually include 4,170 sites that would cost about $40 billion to remedy.
36. For example, the Occupational Safety and Health Administration (OSHA) conducts safety inspections in 11,000 firms nationwide. In fiscal year 1983, OSHA inspected 2.5 percent of all private firms. This included 15.3 percent of all manufacturing plants (SIC 20–30), 21.6 percent of all chemical manufacturers, and 28.3 percent of all industrial chemicals (SIC 286).
37. U.S. Bureau of Labor Statistics, *Occupational Injuries and Illnesses in the United States by Industry, 1983*, U.S. Bureau of Labor Statistics Bulletin 2236 (Washington, D.C.: Department of Labor, June 1985.
38. World Environment Center, *The World Environment Handbook* (New York: World Environment Center, 1984).
39. P.G. Mathai, "Belated Awakening," *India Today*, January 31, 1985, pp. 112–13.

40. H.J. Leonard, "Pollution and Multinational Corporations in Rapidly Indus-trializing Nations" (Research paper, the Conservation Foundation, Washing-ton, D.C., December 1984).
41. Lipton, *The Urban Bias*.
42. United Nations Center for Human Settlements (HABITAT), *The Residential Circumstances of the Urban Poor in Developing Countries* (New York: Praeger Publishers, 1981).

Chapter 3

1. In 1984 the company earned an after-tax profit of $6.80 (per $100 share invest-ment)—well below the industry average of $11.50 and also below its rivals, such as Dow Chemicals ($10.70), Du Pont ($12.00), and Monsanto ($11.80) (M.A. Hiltzik, "Carbide Has Long History of Difficulty," *Los Angeles Times*, 19 August 1985).
2. "Union Carbide: Its Six-business Strategy Is Tight on Chemicals," *Business Week*, September 24, 1979, p. 93.
3. Union Carbide Corporation, *Annual Report* (Danbury, Conn.: Union Carbide Corporation, 1984).
4. It reported a before-tax profit of Rs 148 million, and a profit after tax and In-vestment Allowance Reserve of Rs 87 million. It declared a dividend of Rs 1.50 per share. Net worth per share was Rs 19.02, and earnings per share were Rs 2.86. The company had issued and subscribed share capital of Rs 325.83 million and accumulated reserves and surplus of Rs 293.89 million. Of the 40 million authorized shares of Rs 10 each, 32.58 million were fully paid; of these, 16.58 million shares (50.9 percent) were held by Union Carbide Corpo-ration, U.S.A., the holding company.
5. Union Carbide (India) Ltd., *Annual Reports* (Bombay: Union Carbide (India) Limited, 1979, 1980, 1981, 1982).
6. In 1947, the government set up a Plant Protection Directorate to consolidate knowledge on the methods of safeguarding agricultural crops, almost 30 per-cent of which were annually destroyed by pests. In the early days, pesticides were used mostly by the government for public health purposes. DDT and BHC were commonly used to control malaria and other diseases transmitted by insect vectors. In 1956, plant protection measures using pesticides cov-ered only 2.4 million hectares. The area under chemical pest control had doubled by 1980.
7. Barriers to entry into the industry were low because of the modular nature of production technology. Pesticides could be "formulated" by mixing inter-mediate chemical compounds and packaging them. This required modest working capital and low fixed capital for the relatively inexpensive mixing machinery necessary.
8. K.C. Dhingra, *Handbook of Pesticides* (New Delhi: Small Industries Research Institute, 1978).
9. Several factors contributed to this decline. First, farmers did not possess the required skills and technical know-how to use pesticides effectively. They needed both technical training and attitude changes to adopt pesticides and

other innovative farming methods successfully. Second, individual farmers discovered they could not fight insect infestation alone. Pests would travel from sprayed farms to unsprayed neighboring farms and back again as the original spray wore off. This substantially reduced pesticide effectiveness. Third, pesticide expenditures received third priority, after seeds and fertilizers. In addition, they were bought only if crops had been large and thus worth the expense. Good crops depended in part on the monsoons. This made the pesticides market cyclical, volatile, and uncertain.

10. Government of India, *Economic Survey 1983–84* (New Delhi: Government of India Press, 1984).

11. Town and Country Planning Department, *Bhopal Development Plan* (Bhopal: Municipal Corporation, 1975).

12. Chemical reactions involved in the production of MIC are as follows:
$$2C + O_2 = 2CO$$
$$CO + Cl_2 = COCl_2$$
$$COCl_2 + CH_3NH_2 = CH_3NHCOCl + HCl$$
$$CH_3NHCOCl \rightarrow CH_3NCO + HCl$$

13. The tanks were made of SS 304/SS 316 steel, with a nominal diameter of 2,400 mm (8 feet) and nominal length of 12,000 mm (40 feet). They were designed for full vacuum to 2.72 kg/sq.ca.g (40 psig) at 121°C. Two tanks could hold approximately 90 tons of MIC, which is sufficient for 30 days of Sevin production using stored MIC.

14. Union Carbide (India) Ltd., *Operating Manual Part II Methyl Isocyanate Unit* (Bhopal: Union Carbide (India) Ltd., February 1979).

15. S. Diamond, "The Bhopal Disaster: How It Happened," *New York Times*, 28 January 1985; S. Diamond, "The Disaster in Bhopal: Workers Recall Horror," *New York Times*, 30 January, 1985.

16. International Confederation of Free Trade Unions, *The Trade Union Report on Bhopal* (Geneva: International Confederation of Free Trade Unions, July 1985).

17. Union Carbide Corporation, *Bhopal Methyl Isocyanate Incident Investigation Team Report* (Danbury, Conn.: Union Carbide Corporation, March 1985).

18. Union Carbide Corporation, *Operating Safety Survey CO/MIC/Sevin Units Union Carbide India Ltd. Bhopal Plant* (Danbury, Conn.: Union Carbide Corporation, May 1982).

19. S. Adler, "Carbide Plays Hardball in Court," *American Lawyer*, November 1985.

20. U.S. District Court, Southern District of New York, "Memorandum of law in opposition to Union Carbide Corporation's Motion to dismiss these actions on the grounds of forum and non convenience." MDL Docket No. 626 Misc. No. 21–38 (JFK) 85 Civ. 2696 (JFK). December 6, 1985.

21. Union Carbide Corporation, *Operational Safety Survey*.

22. J. Mukund, *Action Plan-Operational Safety Survey May 1982* (Bhopal: Union Carbide (India) Ltd., 1982).

23. Diamond, "The Bhopal Disaster."

24. C. Perrow, *Normal Accident: Living with High Risk Technologies* (New York: Basic Books, 1984).

25. E. Munoz, Affidavit to the Judicial Panel on Multidistrict Litigation. Federal Court of New York: Southern District, New York, January 28, 1985.

26. P. Bidwai, "Plant Design Badly Flawed," *The Times of India,* 27 December 1984, p. 1; B. Bowonder, J.X. Kasperson, and R. Kasperson, "Avoiding Future Bhopals," *Environment* 27, no. 7 (1985): 6–37.

27. Union Carbide Corporation, *Bhopal Methyl Isocyanate Incident Investigation Team Report* (Danbury, CT: Union Carbide Corporation, 1985); S. Varadarajan et al., *Report on Scientific Studies in the Factors Related to Bhopal Toxic Gas Leakage* (New Delhi: Council of Scientific and Industrial Research, December 1985).

28. Analysis of representative core samples from tank E610 showed the following compounds: methylisocyanate trimer (MICT) 55.71 percent; dimethylisocyanurate (DMI) 21.42 percent; chloride 4.33 percent; trimethyl amine (TMA) 3.384 percent; dione 3.13 percent; dimethyl amine (DMA) 1.978 percent; trimethyl urea (TMU) 1.52 percent; dimethyl urea (DMU) 1.29 percent; monomethyl amine (MMA) 1.02 percent; trimethyl biuret (TMB) 0.94 percent; and tetramethyl biuret (TRMB) in traces.

In addition, the following metallic ions were found: iron (Fe) 1,275 ppm; chromium (Cr) 260 ppm; nickel (Ni) 95 ppm; sodium (Na) 60 ppm; cadmium (Cd) 20 ppm; and magnesium (Mg) 3 ppm.

29. The history of Bhopal dates back to around 1010 A.D., when a Hindu king named Raja Bhoj established the town on a picturesque site with a lake and rolling hills. He also created a second lake, now known as the Upper Lake, by constructing an earthen dam. This lake still serves as the main source of drinking water for the town. Little is known about this early period of Bhopal's history except that for many years it was the battleground for local feuding rulers.

In the early 1700s, Rani (Queen) Kamlavati ruled Bhopal under the protection of the emperor of India, Aurangazeb. After the emperor's death, she invited Dost Mohammed Khan, a muslim chieftain from a neighboring town, to be the protector of her territory. But after her death Dost Mohammed Khan annexed Bhopal to his own kingdom, fortified it, and built it up as a capital of a feudal state. The decline of the Moghul Empire engendered a period of struggle during which local chieftains fought to establish control over land and establish independent kingdoms.

By the mid-nineteenth century, the East India Company had established its rule over large parts of India and was entering into protection pacts with local chieftains. In 1881 the Bhopal Nawab (moslem ruler) Nazar Mohammed Khan entered into a protection pact with the company government. From then until 1947, when India gained independence, the city state was ruled by successive begums and nawabs.

This change of rulers from Hindus to Muslims, and the postindependence establishment of a secular democratic state, gave Bhopal some unique cultural and social characteristics. The city has a large population of muslims in an officially secular but predominantly Hindu country. In the preindependence period, the local language was Urdu (a combination of Hindi and Arabic). Even today, although the official languages are Hindi and English, the local dialect, called "Bhopali," has a heavy Urdu influence.

30. The contribution of the officially registered manufacturing sector to the State Domestic Product at constant prices was 4.8 percent in 1970–71 and 6.5 per-

cent in 1978–79, and of the unregistered sector was 4.6 percent and 4.7 percent, respectively (A.C. Minocha, "Changing Industrial Structure of Madhya Pradesh: 1960–1975," *Margin* 4, no. 1 (1981): 46–61).

31. The exception to this was the brief Janata party rule after the period of emergency in 1977, when Indira Gandhi was ousted from power. Ever since her return to power, Congress (I) has been the dominant political force in the state.

32. This public-sector undertaking was at that time the largest facility of its kind in Asia, manufacturing sophisticated thermal and hydroelectric power generation and transmission equipment in collaboration with several foreign companies.

33. Minocha, "Changing Industrial Structure of Madhya Pradesh: 1960–1975."

34. More than 75 percent of all towns in Madhya Pradesh have populations of less than 20,000 each. In the past decade, some parts of the state, especially the Malwa region (in which Bhopal is located), have experienced rapid urbanization. The number of towns in the state has grown from 232 in 1971 to 303 in 1981, and the decinnial growth rate in the region over the decade 1971–81 has been 56.07 percent in contrast to 38.60 percent for the country as a whole. In 1981, 47.0 percent of the urban population lived in 14 Class I towns (population over 100,000), 18.0 percent lived in 28 intermediate towns (population of 50,000 to 100,000), and 35.1 percent lived in 261 small towns (population less than 50,000). Town and Country Planning Department, *Bhopal Development Plan*.

35. A.C. Minocha and J. Yadav, "Urbanization in Madhya Pradesh" (Unpublished manuscript, Bhopal University, 1984).

36. National Productivity Council, *Industrial Air Pollution: Problems and Control*. National Seminar Proceedings, April 16–17, 1981 (New Delhi: National Productivity Council, 1981).

37. World Environment Center, *The World Environment Handbook* (New York: World Environment Center, 1984).

38. S. Hazarika, "India Journalist Offered Warning," *New York Times*, 11 December 1984; "Probe Report Gathered Dust for 3 Years," *The Times of India*, 2 January 1985.

Chapter 4

1. These include the Royal Commonwealth Society for the Prevention of Blindness, which founded the Bhopal Eye Hospital; the Self-Employed Women's Association (SEWA), which started a medical clinic, a children's school, and a craft center for women; and Action for Gas Affected People (AGAPE), which established a medical clinic with doctors and X-ray facilities, and started an employment program for victims.

2. This set includes the Zaherili Gas Kand Sangarsh Morcha (Poisonous Gas Episode Struggle Front), the Nagrik Raahat aur Punarvas Samiti (Citizens Committee on Relief and Rehabilitation), the Delhi Committee on Bhopal Gas Disaster, and the Trade Union Relief Front. They were helped by existing professional volunteer groups, such as the Medico Friends Circle, the

Delhi Science Forum, the Voluntary Health Association of India, the Society for Participatory Research in Asia, the Center for Science and Environment, Kerala Sahitya Shastra Parishad, the People's Union for Civil Liberties, the People's Union for Democratic Rights, and the Lawyers Collective.

3. People's Union for Democratic Rights, *Who Are the Guilty: Special Report* (New Delhi: People's Union for Democratic Rights and People's Union for Civil Liberties, 1984).

4. One year after the accident occurred, the government reassessed the death toll. The number of deaths was revised from 1756 to 1773 to include deaths that occurred in neighboring districts and states ("Call to Ascertain Total Casualties," *Madhya Pradesh Chronicle*, 28 December 1985). Eight months later the government again revised the death toll to 2,200.

5. Government of Madhya Pradesh, "Medical Treatment and Arrangement Made for the Affected Cases of Poisonous Gas which Leaked Out from Union Carbide Factory on December 2–3, 1984" (Unpublished document prepared by the Directorate of Medical Services, December 1984).

6. An Indian government official in a public meeting organized by the Citizen's Commission on Bhopal, in New York, stated that though the official count was still 1,754, the government estimated it could go up to 2,000.

7. P. Boffey, "Few Lasting Effects Found among Indian Gas-leak Survivors," *New York Times*, 20 December 1984; Indian Council of Medical Research, "The Bhopal Disaster—Current Status (The First Nine Days) and Programme of Research" (Bhopal: ICMR, December 11, 1984); W.K. Stevens, "Encouraging Prognosis Is Seen for Gas Victims," *New York Times*, 15 December 1984.

8. Several studies in the toxicology literature had documented immediate effects of MIC (H.F. Smyth, "Current Status of Knowledge about the Toxicity of Methyl Isocyanate" [Unpublished paper, Mellon Institute, Carnegie-Mellon University, January 1980]). Tests on rabbits indicated that skin was necrosed by contact with undiluted MIC and eyes were severely damaged (Mellon Institute, "Range Finding Tests on Methyl Isocyanate" [Mellon Institute Special Report 26-75, Carnegie-Mellon University, 1963]). It was known to be an irritant to humans at low vapor concentrations, a potent skin sensitizer, an immune suppressant, and a cross-sensitizing agent (H.F. Smyth et al., *Methyl Isocyanate: Acute Inhalation Toxicity, Human Responses to Low Concentration, Guinea Pig Sensitization, and Cross Sensitization to Other Isocyanate*, Mellon Institute Special Report 33-19 [Pittsburgh: Carnegie-Mellon University, 1970]; G. Kimmerle and A. Eben, "Toxicity of Methyl Isocyanate and How to Determine Its Quantity in Air," *Archives of Forensic Toxicology* 20 [1964]: 235–41).

Studies done on rats, mice, and guinea pigs found this chemical to be highly toxic by all routes of administration (D.E. Dodd, L.C. Longo, and D.L. Eisler, "Methyl Isocyanate: Six-hour LC-50 Vapour Inhalation Study on Rats, Mice and Guinea Pigs" [Project report 45-62, Bushy Run Research Center, Export, Penn., Union Carbide Corporation, 1982, chap. 4, p. 26]; L.C. Longo and D.E. Dodd, "Methyl Isocyanate Eight-day Vapour Inhalation Study" [Project report 43-122, Bushy Run Research Center, Export, Penn., January 1981]; U.C. Pozzani and E.R. Kinkead, "Animal and Human Response to Methyl Isocyanate" [Paper delivered at the meeting of the American Industrial Hygiene Association, Pittsburgh, Penn., May 16–20, 1966]). The few studies done with human subjects provided little data. Pozzani and Kinkead ("Animal and Human Response to Methyl Isocyanate") exposed seven human subjects to an approximate concentration of one ppm of MIC vapor for ten minutes. All subjects reported eye irritation within four minutes, and three subjects reported nose and throat irritation after the ten-minute exposure.

9. The by-products included monomethylamine, 1, 3- dimethylurea (DMU), 1, 3, 5- trimethylbiuret (TMB), ammonium chloride, methyl-substituted amine hydrochlorides (DMA-HCl and TMA-HClO, monomethyl urea, trimethyl-urea, MIC trimer dimethylisocynate (DMI), dione, 1, 1, 3, 5- tetramethylbiuret chloroform, etc. (Union Carbide Corporation, *Bhopal Methyl Isocyanate Incident Investigation Team Report* [Danbury, Conn.: Union Carbide Corporation, March 1985]). Traces of cyanide were also detected in a 1985 study done by the Central Water and Air Pollution Control Board (*Report of the Central Water and Air Pollution Control Board (Gas leak episode at Bhopal),* in *The Bhopal Tragedy—One Year After,* An APPEN Report [Malasia: Sahabat Alam, 1986]).

10. S.R. Kamath et al., "Preliminary Observations on Early Toxicity in Subjects Exposed to the Isocyanate Gas Leak Disaster at Bhopal," *Journal of Post Graduate Medicine* 31, no. 2 (1985): pp. 63–72; Medico Friends Circle, *The Bhopal Disaster Aftermath: An Epidemiological and Socio-Medical Survey* (Bangalore, India: Medico Friends Circle, October 1985).

11. Government of Madhya Pradesh, *Bhopal Gas Tragedy, Relief and Rehabilitation—Current Status* (Bhopal: Government of Madhya Pradesh, August 1985), p. 57.

12. Arun Subramaniam, "The Dangers of Diagnostic Delay," *Business India,* August 12–25, 1985, pp. 128–133; Olga Tellis, "The Crime Continues," *Sunday,* April 7–13, 1935, pp. 23–29; "Serious Blood Disorder by MIC Exposure," *Madhya Pradesh Chronicle,* 10 November 1985; S. Weisman, "Medical Problems Continue in Bhopal," *New York Times,* 31 March 1985.

13. These compounds are described in note 9 of this chapter.

14. There are two conditions under which cyanide could have been generated during the accident. First, pure MIC in the absence of chloroform heated to 300°C produces hydrogen cyanide. Second, breakdown products of MIC include MIC trimer (MICT), dimethylisocyanurate (DMI), dimethyl urea (DMU), dione, and trimethyl biuret (TMB). These by-products heated in the presence of MIC release traces of hydrogen cyanide. These reactions have been confirmed in laboratory experiments (S. Varadarajan et al., *Report on Scientific Studies in the Factors Related to Bhopal Toxic Gas Leakage* [New Delhi: Council of Scientific and Industrial Research, December 1985]). Both of these conditions were possible in the accident.

15. Medico Friends Circle, *The Bhopal Disaster Aftermath.*

16. S. Khandekar, "Painful Indecision," *India Today,* January 31, 1985, pp. 84–85.

17. *Report of the Central Water and Air Pollution Control Board,* in *The Bhopal Tragedy—One Year After,* 1986.

18. T. Narayan et al., "Rationale for the Use of Sodium Thiosulphate as an Antidote in the Treatment of the Victims of the Bhopal Gas Disaster—A Review" (Unpublished paper, Medico Friends Circle, Bangalore, India, June 1985); S. Weisman, "Doctors in India Disagree on Drug," *New York Times,* 10 April 1985.

19. Union Carbide Corporation, *New Light on Bhopal Cyanide Controversy: Toxicology, Treatment and Third-Party Analysis* (Danbury, Conn.: Union Carbide Corporation, May 1985).

20. Khandekar, "Painful Indecision."

21. A. De Grazia, *A Cloud over Bhopal* (Bombay: Kalos Foundation, 1985); W. Morehouse and A. Subramaniam, *The Bhopal Tragedy: What Really Happened and What It Means for American Workers and Communities at Risk* (New York: Council on International Public Affairs, 1986).

22. S. Diamond, "The Disaster in Bhopal: Lessons for the Future," *New York Times*, 3 February 1095.

23. Personal interviews revealed that many families in affected neighborhoods lived hand-to-mouth to begin with. They worked during the day to be able to afford to eat at night.

24. S. Vishwanathan and R. Kothari, "Bhopal: The Imagination of a Disaster," *Lokayan* 3 (1985): 48–75.

25. "2nd Evacuation," *Times of India*, 18 December 1984.

26. J. Bates, "Fear Is Residue of Bhopal Tragedy," *APA Monitor*, July 1985, p. 15; "Psychological Effects on Gas Victims?" *Madhya Pradesh Chronicle*, May 21, 1985.

   Psychologists and social workers from King George Medical College, Lucknow, along with local doctors, did a pilot survey of mental conditions in affected areas. The study of 393 families, involving 1,195 subjects, revealed psychiatric morbidity in 193 subjects. A majority of them suffered from neurotic disorders, such as neurotic depression (66.3 percent) and anxiety (33.5 percent) (Government of Madhya Pradesh, *Bhopal Gas Tragedy Relief and Rehabilitation—Current Status* [Bhopal: Government of Madhya Pradesh, August 1985]).

27. "Bhopal—How Women Suffered," *Manushi* 29 (1985): 36–37.

28. Medico Friends Circle, *The Bhopal Disaster Aftermath*.

29. "Gas Disaster Results in 400 Abortions," *Madhya Pradesh Chronicle*, 4 October 1985.

30. P. Prakash, "Neglect of Women's Health and Issues," *Economic and Political Weekly*, December 14, 1985, p. 219b.

31. Routine hematological findings in four affected buffaloes revealed no significant cellular changes in morphology of erythroytes and leucocytes. Hemoglobin values varied from 11.8 to 14.8 g percent, which was normal; the leucocytic count varied within a normal range of 11,500 to 12,000 per cubic millimeter of blood. There was elevation in blood urea nitrogen (BUN). Choline esterage activity observed in several animals suggested damage to the central nervous system (Indian Council of Agricultural Research, *The Bhopal Disaster Effect of MIC Gas on Crops, Animals and Fish* [New Delhi: Indian Council of Medical Research, 1985]).

32. Twelve plants were severely affected, seventeen were partially affected, and six were less affected. Severely affected plants were those in which either the leaves had been completely shed or more than 50 percent of the leaves bore signs of being burned or had turned black. Partially affected plants were those in which 25 to 50 percent of plant foliage had burned or turned black but where leaf shedding was not pronounced. Less affected plants were those

in which 10 to 25 percent of leaves had been affected by necrosis (black or white spots).

33. R. Prasad and R.K. Pandey, "Methyl Iso-cyanate (MIC) Hazard to the Vegetation in Bhopal," *Journal of Tropical Forestry* 1, no. 1 (1985): 40–50.

34. Indian Council of Agricultural Research, *The Bhopal Disaster Effect of MIC Gas on Crops, Animals and Fish.*

35. UCC received much negative publicity through this event, which was rated as the third most important media story of the year. Both the *New York Times* and the *Times of India* ran front page stories on Bhopal every day for two weeks.

36. J. Browning, Press release by Union Carbide Corporation, Danbury, Conn., March 20, 1985.

37. Environmental Protection Agency, *Multi-Media Compliance Inspection Union Carbide Corporation, Institute, West Virginia* (Philadelphia: EPA-RIII, Environmental Services Division, 1985). Actually, there were ninety-seven MIC leaks during this period, but only twenty-eight were large enough to warrant reporting under existing laws.

38. G. Anders, "Carbide's Destiny Shaped by Holders," *Wall Street Journal*, 7 January 1986; B. Meier and J.B. Stewart, "A Year after Bhopal, Union Carbide Faces a Slew of Problems," *Wall Street Journal*, 26 November 1985, p. 1.

39. A.J. Broder, "Trial Tactics, Techniques," *New York Law Journal* (January 10, 1985): 1–3.

40. S. Khandekar, "An Area of Darkness," *India Today*, June 30, 1985, pp. 134–36.

41. The exact figure of offer made by Union Carbide is not known, but is has been quoted at between $180 and $300 million. A *Business Week* story quoted the Carbide offer as $60 million immediately plus $180 million over the next thirty years (April 22, 1985, p. 38). A *Wall Street Journal* article stated that Carbide's offer was for $100 million in cash to finance some $230 million over thirty years (November 26, 1985).

42. The abrupt breakdown of out-of-court negotiations was surprising. There were two speculations about its cause. Some observers saw it as the government yielding to pressure from activists and the media in India to continue legal proceedings against UCC and not to settle "cheaply." Settlement at around $200 million, without investigations into causes of the accident, was considered by activists to be a "sell out" to Union Carbide (R. Hager, "Taking Carbide to Court," *Business India*, March 25–April 7, 1985) pp. 130–137. The second speculation was more sinister. There were rumors that UCC was exerting pressure for early, low-cost settlement by influencing a key government negotiator. When senior officials discovered this plot, the negotiations were abandoned.

43. J. Nelson-Horchler, "Fallout from Bhopal," *Industry Week*, May 13, 1985, pp. 44–48.

44. By July 1985, 160 of the 180 CMA member firms had appointed a CAER coordinator, and over 1,200 people were registered for workshops. By July 1986, 1020 plant sites out of 1590 participation sites had formed CAER coordinating groups.

45. J. Nolan, "Bhopal Likely to Alter Drastically How Insurance Is Written Abroad," *Journal of Commerce* (December 31, 1984): 1.

46. In the U.S. Congress the following bills were introduced:

- *HR 966*, March 1, 1985, to amend the Hazardous Materials Transportation Act to provide for establishment of regional training centers to assist in improving the emergency response and enforcement capabilities of state and local personnel.
- *HR 1105*, February 19, 1985, to amend the Hazardous Materials Transportation Act to restrict the transportation of radioactive material through large cities.
- *HR 1660*, March 21, 1985, to amend the Solid Waste Disposal Act and the Toxic Substances Control Act to prevent releases of toxic and hazardous substances that are presently not adequately controlled, to establish a community right to know.
- *HR 2118*, April 18, 1985, amending chapter 18 of title 10, United States Code to establish a toll-free telephone service from which emergency assistance information may be obtained in case of an accident.
- *HR 2576*, May 22, 1985, to control toxic releases into the air, and for other purposes.

47. The "No More Bhopal Network" was formed by the International Organization of Consumers Union, Malaysia, and the Japan Bhopal Monitoring Group was formed in Japan. In the United States, the following groups were formed to help Bhopal victims and examine the implications of Bhopal for U.S. communities: the National Bhopal Disaster Relief Foundation, Inc., Miami, Florida; the Citizens Commission on Bhopal, New York; and People Concerned About MIC, Institute, West Virginia. Among the public and international organizations that initiated studies of these issues were: Environment Canada, Ontario; the Environmental Protection Agency, Washington, D.C.; the Industrial Crisis Institute, New York; the National Institute of Occupational Safety and Health, Washington, D.C.; the World Health Organization, Geneva; and the Center for Transnational Corporations, New York. More than half a dozen documentary films were made by the British Broadcasting Corporation, the Canadian Broadcasting Corporation, Granada TV, the Highlander Center, and Indian filmmakers Tapan Bose and Suhasini Mulay.

Chapter 5

1. R.P. Gephart, "Making Sense of Organizationally Based Environmental Disasters," *Journal of Management* 10, no. 2 (1984): 205–25; H. Molotch and M. Lester, "Accidental News: The Great Oil Spill," *American Journal of Sociology* 81 (1975): 235–60.
2. G.T. Allison, *Essence of Decision: Explaining the Cuban Missile Crisis* (Boston: Little Brown, 1971).
3. Gephart, "Making Sense of Organizationally Based Environmental Disasters."
4. K. Weick, *The Social Psychology of Organizing* (Reading, Mass.: Addison-Wesley, 1979).
5. P. Shrivastava and S. Schneider, "Organizational Frames of Reference," *Human Relations* 37, no. 10 (1984): 795–809.

6. W.H. Starbuck and B.L.T. Hedberg, "Saving an Organization from a Stagnating Environment," in *Strategy & Structure=Performance*, ed. H.B. Thorelli (Bloomington: Indiana University Press, 1979), p. 253.

7. Data on accident-related deaths are extremely sparse and unreliable. Although the Central Statistical Organization quoted figures for 1977 and 1978 of 451 and 471 deaths by factory mishaps, and 10,207 and 11,117 deaths by railway accidents, these figures are generally believed to be underestimated. Recent estimates by labor groups claim as many as 35,000 deaths by accidents in government establishments (Sagar Dhara, paper delivered at the conference on "The Bhopal Tragedy: Its Implications for Developing and Developed Countries," organized by the Workers Policy Project and the Labor Institute, New York, March 1985).

8. Director of Information and Publicity, "We Shall Overcome" (Brochure, Directorate of Information and Publicity, Bhopal, November 1985).

9. "Call to Ascertain Total Casualties," *Madhya Pradesh Chronicle*, 28 December 1985.

10. M. Nanda, "Secrecy Was Bhopal's Real Disaster," *Science for the People* 17, no. 6 (1985): 12–17.

11. On December 11, 1984, group meetings were held with management and staff. Managers were appointed to meet groups of staff members, answer their questions, and maintain communications. As time passed, these meetings became regular. The initial meetings were attended by about 160 staff members; subsequently, attendance declined sharply.

12. D. Thakore, "Interview with V.P. Gokhale," *Business India*, May 27, 1985.

13. W. Anderson, "More on Bhopal from Warren Anderson," Letter to the Editor, *Business Week*, January 20, 1986, p. 8.

14. UCC later made smaller contributions to several third parties to do relief work in Bhopal. In distributed $50,000 among three organizations: the Missionaries of Charity, New York; the Share and Care Foundation, New Jersey; and the Lions Club International Foundation, Illinois. In addition, it authorized an expenditure of $1.1 million for studies and visits of medical experts to and from Bhopal ("Carbide's Restructuring Plan," *Hexagon* [Union Carbide in-house magazine], 1985).

15. S. Adler, "Carbide Plays Hardball in Court," *American Lawyer*, November 1985; R. Mokhiber, "Paying for Bhopal: Union Carbide's Campaign to Limit Its Liability," *Multinational Monitor* 6, no. 10 (1985): 2–5.

16. J. Browning, Press release from Union Carbide Corporation, Danbury, Conn., March 20, 1985.

17. In August 1985 UCC President Alec Flamm himself visited IDS Financial Services in Minneapolis, a holder of 1 percent of Carbide stock, to brief IDS on his company's prospects (G. Anders, "Carbide's Destiny Shaped by Holders," *Wall Street Journal*, 7 January 1986.

18. "Seven-Step Restructuring Plan Follows Reorganization," *Union Carbide World*, November–December 1985, pp. 8–9.

19. S. Vishwanathan and R. Kothari, "Bhopal: The Imagination of a Disaster," *Lokayan* 3 (1985): 48–75.

Chapter 6

1. H. Smets, "Compensation for Exceptional Environmental Damage Caused by Industrial Activities" (Paper delivered at the conference on Transportation, Storage and Disposal of Hazardous Materials, IIASA, Laxenburg, Austria, July 1–5, 1985).
2. T.N. Gladwin, "The Management of Environmental Conflict: A Survey of Research Approaches and Priorities" (Paper delivered at the workshop/conference on Environmental Meditation: An Effective Alternative, Reston, Virginia, January 11–13, 1978).
3. E.S. Muskie and D.J. Greenwald, "An Overview of the Nestle Infant Formula Audit Commission: Is It a Model," *Journal of Business Strategy* (Spring 1986): 19–23.
4. In centrally coordinated economies, the choice of technologies is made through national economic plans. In other economies, the choice is decentralized and controlled by corporations. Developing countries, lacking resources, have more limitations on the choice of technologies.
5. R.W. Kates, ed., *Managing Technological Hazard: Research Needs and Opportunities*, Monograph #25 (Boulder, Colo.: Institute of Behavioral Science, University of Colorado, 1977).
6. This directive requires manufacturers to take all necessary measures to prevent accidents during plant design, construction, and operation. It also requires them to anticipate possible causes of accidents, monitor critical points in the production process, introduce stringent safety measures, and adopt emergency plans for limiting the effects of accidents. The manufacturers are also required to provide the government with a wide range of information on the plant's hazards.

    Manufacturers must provide the government information on certain hazardous substances, when storage exceeds designated amounts; the number of people working on the site; technological processes; safety measures; arrangements for dealing with malfunctions and emergencies; details of emergency plans and equipment available for on-site use; and arrangements for initiating emergency plans.

    Just as important, the directive imposes an obligation on the European nations to create an environmental department that collects all this information, monitors it, and establishes emergency plans for off-site use. In addition, the public and neighboring countries must be provided with information about safety measures and emergency procedures (E. Lykke, "Avoiding and Managing Environmental Damage from Hazardous Industrial Accidents," in *Avoiding and Managing Environmental Damage from Major Industrial Accidents* [Pittsburgh: Air Pollution Control Association, 1986]).

    The limited role of the public is a key shortcoming of the Seveso Directive. There is insufficient public information available. As one analyst of the directive suggested, "What people really want is to know enough so that they can develop their own perceptions of whether the risk [posed by a facility] is acceptable or not (Lykke, "Avoiding and Managing Environmental Damage,"

p. 43). For these perceptions to be formed realistically, citizens need far more information as well as resources that will empower them to use it. The information provided to governments could easily be shared with the public, and a decentralized system of risk management could be established that would provide individual communities with the resources to assess risks and act on their assessments.

Despite its shortcomings, the Seveso Directive is a clear improvement over the fragmented, ad hoc regulatory responses to individual crises that we have witnessed in the past.

7. R. Batstone, "Experience in the Application of Hazard Assessment on World Bank Projects," in *Avoiding and Managing Environmental Damage from Major Industrial Accidents* (Pittsburgh, Penn.: Air Pollution Control Association, 1986). Environmental considerations are introduced at the project identification stage. Those projects that will have significant environmental effects are singled out for review. The projects are subject to reviews, investigations, environmental cost-benefit analysis, and analysis of measures taken to mitigate serious adverse effects. During project preparation, the bank assists borrowers in conducting environmental inquiries and identifying problem areas. Later, during project appraisal, preventive measures are incorporated into the design and operation of the projects. Environmental requirements are incorporated into the loan agreement. Implementation of the projects includes implementing the environmental measures. Finally, environmental post-audits are conducted to assess the effectiveness of preventive safety measures and the need for further action (World Bank, "Guidelines for Identifying, Analyzing and Controlling Major Hazard Installations in Developing Countries" [Washington, D.C.: World Bank, 1985]).

Other procedures used by the Bank are similar to the Seveso Directive, in that they require corporations and governments to demonstrate that major hazards have been recognized, that information about key variables has been exchanged, and that measures have been taken to prevent and minimize the consequences if an accident should occur. These measures include reduction in inventories of flammable and acutely toxic substances; redesign of production processes to eliminate the need for storage; use of double integrity containment; installation of automatic shut-down systems; improved emergency systems and evacuation planning; public alert systems; creation of buffer zones and modified siting of plants; and continuous hazard monitoring assessment (World Bank, *Manual of Industrial Hazard Assessment Techniques* [Washington, D.C.: World Bank, 1985]).

8. Even if the buffer zone is correctly specified, often, as in Bhopal, it is not adhered to. Both workers and businesses have an economic interest in being in close proximity to each other, particularly in developing countries, where workers must save on transportation costs. In large Third World cities—Singapore, Mexico City, Bombay, Lagos, Caracas, Sao Paolo—the presence of shantytowns next to industrial plants is the rule rather than the exception. This proximity has led to tragedy in many industrial accidents. For example, gasoline leaking from pipelines is an everyday occurrence in petrochemical plants.

It caused concern in Sao Paolo in 1984 only because it occurred in a shanty-town, setting off a fire that swept through the area and killed 500 people.

Once such conditions are created, it is virtually impossible to change them. Alternative housing is enormously expensive, and even if it were available the shantytown residents would resist moving because of well-engrained so-cial habits, customs, and the convenience of a prolonged tenure in one loca-tion. Most governments simply ignore the shantytowns or, as in the case of Bhopal, legalize them in hopes of discouraging threatening "landlords" who extort rents from the residents. Even if the residents cannot or will not be moved, the risks to them can be reduced if they are provided information about the nature of the hazard nearby and emergency procedures to follow in case of an accident.

9. New Jersey State Act 13:1E-49.
10. In essence, this would mean shifting the responsibility for the negative ex-ternalities of production (environmental pollution, occupational hazards, and accidents) from the government to the industries themselves.

In the United States, many local governments require real estate develop-ers to pay for external infrastructure, such as roads, sewers, and parks. In developing countries, a similar approach would make sense because corpo-rations—often more than the government—possess the know-how and the resources to deal with the externalities. The infrastructure needed to deal with the immediate negative effects of industrial plants should be made the responsibility of the plant operators.
11. These failures come about for many reasons. Often the government is unable to monitor performance and punish offenders. Sometimes the judicial sys-tem cannot prove offenses and enforce deterring punishment. Just as often, policy implementers disagree with the policies they are working with, en-forcement is purposely lax, or loopholes allow many businesses to evade reg-ulations (J.L. Pressman and A. Wildavsky, *Implementation*, 3d ed. [Berkeley, Calif.: University of California Press, 1984]).
12. M. Grindle, *Politics and Policy Implementation in the Third World* (Princeton, N.J.: Princeton University Press, 1980).
13. Richard G. Vetter, "An Expert Systems Application for Crisis Management," *International Industrial Engineering Conference Proceedings* (Norcross, Ga.: Insti-tute of Industrial Engineers, 1986). Obviously, this recommendation is not realistic unless there exists a compatible system for linking these databases. The author is presently working on such a project.
14. R. Grosse, "Codes of Conduct for Multinational Enterprise," *Journal of World Trade Law* (1982); C.S. Pearson, *Down to Business*, World Resources Institute Study #2 (Washington, D.C.: World Resources Institute, 1985).
15. The International Chamber of Commerce has developed the "Environmental Guidelines for World Industry." The U.N. Center for Transnational Corpo-rations is developing a comprehensive code for environmental protection. Several other codes and guidelines tangentially address plant safety, worker protection, and consumer safety issues (Grosse, "Codes of Conduct for Mul-tinational Enterprise").

16. J. Pearson, *Technology, Environment, and Development* (Washington, D.C.: World Resources Institute, 1984).

17. To be effective, EIAs must comprehensively examine the hazard potential of the proposed facility, options and costs of reducing it, and national and international standards for similar facilities. The analysis must be comprehensive both in terms of life-cycle coverage of the investment and of the many forms of environmental effects. It must include the entire investment cycle—construction, operation, decommissioning, and restoration of the investment site—as well as contingency plans for hazard-related emergencies. It must also assess important effects on the social and cultural environment, such as changes in lifestyles and family and social relationships, enhancement and distortion of local traditions and values, and influences on local politics and the community power structure.

18. P. Shrivastava, "The Unethical Fallout from technical Decisions in Bhopal," *Proceedings of the Sixth Bentley Symposium on Business Ethics* (Worcester, Mass.: Bentley College, 1986).

19. Any emergency plan should include (1) technical and medical assessments of accidents; (2) notification and communication with plant management, local authorities, and neighboring communities; (3) establishment of command and coordination structure and allocation of responsibility of tasks; (4) delineation of protective actions, such as sheltering, evacuation, control of access to contaminated areas, and control of contaminated substances; and (5) support actions to mitigate effects of the accident, such as firefighting, medical treatment, crime prevention, and so on (E.J. Michael, "Elements of Effective Contingency Planning," in *Avoiding and Managing Environmental Damage from Major Industrial Accidents, Proceedings of an International Conference* [Pittsburgh, Penn.: Air Pollution Control Association, 1986]). A broad emergency planning zone should be established, based on the results of the EIA.

20. W.H. Chase, *Issue Management* (Stamford, Conn.: Issue Action Publications, Inc., 1984).

21. Staff members in affected subsidiaries and headquarters require better information. They feel more threatened by the crisis than their counterparts elsewhere, and they often serve as informal spokespersons for the company. Also, customers who are about to place orders, suppliers with whom orders have been placed, and bankers who work with affected subsidiaries must know about the status of their transactions.

    Several vehicles are available to communicate with these various audiences. For communicating with external constituencies, business letters and official notices are frequently used. Some external stakeholders may have specialized communication systems, such as industry newsletters, trade magazines, computer "bulletin boards," and meetings.

22. P. Shrivastava, "A Cultural Analysis of Social Conflicts in Industrial Disasters" (Working Paper, Industrial Crisis Institute, New York, N.Y., May 1985).

23. M. Douglas and A. Wildavsky, *Risk and Culture* (Berkeley: University of California Press, 1982); P. Slovic, B. Fischoff, and S. Lichtenstein, "Cognitive Processes and Societal Risk Taking," in *Cognition and Social Behavior*, ed. J.S. Carroll and J.W. Payne (Potomac, Md.: Erlbaum Associates, 1976); P. Slovic,

B. Fischoff, and S. Lichtenstein, "Rating the Risk," *Environment* 21 (1979): 14–39; O. Renn, "Risk Perception: A Systematic Review of Concepts and Research Results," in *Avoiding and Managing Environmental Damage from Major Industrial Accidents, Proceedings of an International Conference* (Pittsburgh, Penn.: Air Pollution Control Association, 1985).

24. D. Nelkin, *Controversy: Politics of Technical Decisions* (Beverly Hills, Calif.: Sage Publications, 1979).

25. K.T. Erikson, *Everything in Its Path* (New York: Simon and Schuster, 1976).

26. One particularly useful tactic is the creation of a team of leaders. A team brings the resources, contacts, and talents of several people to bear on crisis problems. It provides a vehicle through which individual members can periodically become more or less involved without drastically affecting the group's progress. This flexibility is critical, because most such organizations depend on volunteers. Teams also provide a feeling of solidarity and shared commitment.

27. These included the Zahiri Gas Kand Sangharsh Morcha; Nagrik Rahat aur Punarvas Samiti; the "No More Bhopal Network," formed by the International Union of Consumers Union, Malaysia; the Japan Bhopal Monitoring Group; the National Bhopal Disaster Relief Foundation, Inc., Miami, Florida; the Citizens Commission on Bhopal, New York; and People Concerned About MIC, Institute, West Virginia.

28. A list of international nongovernmental organizations (NGOs) concerned about industrial crisis-related issues is available from the United Nations.

29. A. Barton, *Communities in Disaster* (New York: Anchor Books, 1969). Also see S.P. Sethi, Hamid Etemad, and K.A.N. Luther, "New Sociopolitical Forces: The Globalization of Conflict," *The Journal of Business Strategy* 6, no. 4 (Spring 1986): 24–31.

Suggestions for Further Reading

This list of publications was developed for readers wishing to further explore issues in industrial crises. It is drawn from a larger computerized bibliographic data base developed by the Industrial Crisis Institute, Inc., New York. For further information on the data base contact the Institute at Suite 2B, 100 Bleecker St., New York, NY 10012.

Allee, John S. "Post-sale Obligations of Product Manufacturers." *Journal of Products Liability* 8, no. 2 (1985): 141–66.

Allison, A., A. Carnesale, P. Zigman, and F. DeRosa. *Governance of Nuclear Power.* Report Submitted to the President's Nuclear Safety Oversight Committee. September 1981.

Anderson, W. "Disaster Warning and Communication Processes in Two Communities." *The Journal of Communications* 19, no. 2 (1969):92–104.

An APPEN Report. *The Bhopal Tragedy—One Year After.* Malaysia: Sahabat Alam Malaysia, 1986.

Ashford, N.A. *Crisis in the Workplace: Occupational Disease and Injury.* Cambridge, Mass.: MIT Press, 1976.

Avoiding and Managing Environmental Damage from Major Industrial Accidents. Proceedings of the International Conference. The Air Pollution Control Association; Pittsburgh: 1985.

Bardo, J.W. "Organized Response to Disaster: A Typology of Adaptation and Change." *Mass Emergency* 3 (1978): 87–104.

Barton, Allen H. *Communities in Disaster.* New York: Anchor Books, 1969.

Battisti, F. "Thresholds of Security in Different Societies." *Disasters* no. 4 (1980): 101–105.

Beach, H.D., and F.A. Lucas, eds. *Individual and Group Behavior in a Coal Mine Disaster.* National Academy of Sciences, National Research Council, no. 13. Washington, D.C.: National Academy of Sciences, National Research Council, 1963.

Bem, D.J., M. Wallach, and N. Kogan. "Group Decision Making Under Risk of Aversive Consequences." *Journal of Personality and Social Psychology* no. 1 (1965): 453–60.

Berman, D.M. *Death on the Job.* New York: Monthly Review Press, 1978.

Bettelheim, Bruno. "Individual and Mass Behavior in Extreme Situations." *Journal of Abnormal and Social Psychology* 38 (1943): 417–52.

Billings, Robert S., Thomas W. Milburn, and Mary Lou Schaalman. "A Model of Crisis Perception: A Theoretical and Empirical Analysis." *Administrative Science Quarterly* 25 (1980): 300–16.

Bodenheimer, Tom. "The Malpractice Blow-up." *Health Policy Advisory Center Bulletin* 64 (1975): 12–15.

Boder, David P. "The Impact of Catastrophe." *Journal of Psychology* 38 (1954): 3–50.

Boedecker, Karl A., and Fred W. Morgan. "The Channel Implication of Product Liability Development." *Journal of Retailing* 56, no. 4 (Winter 1980): 59–72.

Bowonder, B., J.X. Kasperson, and R. Kasperson. "Avoiding Future Bhopals." *Environment* 27, no. 7 (1985): 6–37.

Bozeman, Barry, and E. Allen Slusher. "Scarcity and Environmental Stress in Public Organizations: A Conjectural Essay." *Administration and Society* 2 (1978): 335–55.

Broadbent, Donald E. *Decision and Stress.* London: Academic Press, 1971.

Brodeur, Paul. *Outrageous Misconduct: The Asbestos Industry on Trial.* New York: Pantheon, 1985.

Brouillette, J.R., and E.L. Quarantelli. "Types of Patterned Variation in Bureaucratic Adaptations to Organizational Stress." *Social Inquiry* 41 (1971): 39–45.

Brown, L., W.U. Chandler, C. Flavin, C. Pollock, S. Postel, L. Starke, and E.C. Wolf, eds. *State of the World 1985.* New York: W.W. Norton and Co., 1985. (Also *State of the World 1986*, 1986.)

Burton, I. *The Mississauga Evacuation, Final Report.* Toronto: Institute of Environmental Studies, University of Toronto, 1981.

Caldwell, D.F., and C.A. O'Reilly, III. "Responses to Failure: The Effects of Choice and Responsibility on Impression Management." *Academy of Management Journal* 25, no. 1 (1982): 121–36.

Casti, John. *Connectivity, Complexity, and Catastrophe in Large Scale Systems.* London: International Institute for Applied Systems Analysis, 1979.

Castleman, B.I., "The Export of Hazardous Factories to Developing Nations." *International Journal of Health Services* 9 (1979): 569–806.

———. *Asbestos: Medical and Legal Aspects.* New York: Law & Business Inc./Harcourt Brace Jovanovich, Inc., 1986.

Catton, W.R., and R.E. Dunlap. "Environmental Sociology: A New Paradigm." *American Sociologist* 13 (1978): 41–49.

Chase, W.H. *Issue Management: Origins of the Future.* Stamford, Conn.: Issue Action Publication Inc., 1984.

Clark, Richard Charles. *Technological Terrorism.* Old Greenwich, Conn.: The Devin-Adair Company, 1980.

Covello, V.T. "The Perception of Technological Risks: A Literature Review." *Technological Forecasting and Social Change* 23 (1983): 285–97.

Cox. T. *Stress.* Baltimore: University Park Press, 1978.

Dacy, Douglas C., and Howard Kunreuther. *The Economics of Natural Disasters.* New York: The Free Press, 1969.

Danzig, E.R., P.W. Thayer, and Lila P. Gallanter. *The Effects of a Threatening Rumor on a Disaster-ridden Community.* National Academy of Sciences, National Research Council, disaster study no. 10. Washington, D.C.: National Academy of Sciences, National Research Council, 1958.

Davis, M. "A Few Comments on the Political Dimensions of Disasters and Disaster Assistance." *Disasters* 2 (1978): 134–36.

De Grazia, A. *A Cloud Over Bhopal.* Bombay: Kalos Foundation, 1985.

Delhi Science Forum. "Bhopal Gas Tragedy." *Social Scientist* 13, no. 1 (1985): 32–53.

Devine, I. "Organizational Crisis and Individual Response: The Case of the Environmental Protection Agency." Ph.D. diss., Case Western Reserve University, 1983.

Douglas, M., and A. Wildavsky. *Risk and Culture.* Berkeley, Calif.: University of California Press, 1982.

Drabek, T.E. "Methodology of Studying Disasters: Past Patterns and Future Possibilities." *American Behavioral Science* 13 (1970): 332–43.

———. *Human System Responses To Disaster: An Inventory of Sociological Findings.* New York: Springer-Verlag New York Inc., 1986.

Drabek, T.E., J.L. Taminga, T.S. Kilijanek, and C.R. Adams. *Managing Multiorganizational Emergency Responses: Emergency Search and Rescue Networks in Natural Disasters and Remote Settings.* Boulder: Institute of Behavioral Science, University of Colorado, 1981.

Dynes, R.R., A.H. Purcell, D.E. Wenger, P.S. Stern, and R.S. Stallings. *Report of the Emergency Preparedness and Response Task Force, President's Commission on the Accident at Three Mile Island.* Washington, D.C.: U.S. Government Printing Office, 1979.

Dynes, Russell R. *Organized Behavior in Disaster.* Lexington, Mass.: D.C. Heath, 1970.

Environment, Development and Natural Resource Crisis in Asia and the Pacific. Malaysia: Sahabat Alam Malaysia, 1986.

Erickson, K.T. *Everything in its Path.* New York: Simon and Schuster, 1976.

Erman, D.M., and R. Lundman, eds. *Corporate and Governmental Deviance.* New York: Oxford University Press, 1978.

Everest, L. *Behind the Poison Cloud.* Chicago: Banner Press, 1985.

Fink, S. *Crisis Management: Planning for the Inevitable.* New York: AMACON, 1986.

Fritz, C. "Disasters." In *Social Problems,* edited by E.R. Merton and R. Nisbet, 651–94. New York: Harcourt Brace and World, 1961.

Gephart, R.P. "Making Sense of Organizationally Based Environmental Disasters." *Journal of Management* 10, no. 2 (1984): 205–25.

Giddens, A. *Central Problems in Social Theory: Action, Structures and Contradictions in Social Analysis.* Berkeley: University of California Press, 1979.

Gould, L., and C.A. Walker, eds. *Too Hot to Handle: Public Policy Issues in Nuclear Waste Management.* New Haven, Conn.: Yale University Press, 1981.

Green, C.H., and R.A. Brown. *Life Safety: What Is It and How Much Is It Worth.* CP52/78. Borehamwood, Hertfordshire, England: Department of the Environment, Building Research Establishment, 1978.

Grindle, M., ed. *Politics and Policy Implementation in the Third World.* Princeton, N.J.: Princeton University Press, 1980.

Grosse, R. "Codes of Conduct for Multinational Enterprises." *Journal of World Trade Law* (1982).

Guidelines for Identifying, Analysing and Controlling Major Hazard Installations in Developing Countries. Washington, D.C.: The World Bank, 1985.

Habermas, J. *Legitimacy Crisis.* Boston: Beacon, 1975.

―――. *Theory of Communicative Action.* 2 vols. Cambridge, Mass.: MIT Press, 1983.

Hale, A.R., and M. Hale. "Accidents in Perspective." *Occupational Psychology* 44 (1970): 115–21.

Hall, D.T., and R. Mansfield. "Organizational and Individual Responses to External Stress." *Administrative Science Quarterly* 16 (1971): 533–47.

Hamblin, R.L. "Group Integration During a Crisis." *Human Relations* 11 (1958): 67–76.

―――. "Leadership and Crisis." *Sociometry* 21 (1958): 322–35.

Harada, M. "Minamata Disease: Organic Mercury Poisoning Caused by Ingestion of Poisoned Fish." In *Adverse Effects of Floods,* edited by E.F. Patrice Jelliffe and D.B. Jelliffe. New York: Plenum Publishing Corporation, 1982.

Harris, Louis and Associates, Inc. *Risk In a Complex Society.* Chicago: March & McClennon Public Opinion Survey, 1980.

Hartley, R.F. *Management Mistakes.* Columbus, Ohio: Grid Publishing, 1983.

Healy, Richard J. *Emergency and Disaster Planning.* New York: Wiley, 1969.

Hedberg, Bo. "How Organizations Learn and Unlearn." In *The Handbook of Organizational Design,* edited by Paul C. Nystrom and William H. Starbuck, 113–27. New York: Oxford University Press, 1981.

Hermann, Charles F. "Some Consequences of Crisis which Limit the Viability of Organizations." *Administrative Science Quarterly* 8 (1963): 61–82.

Hertzler, J.D. "Crisis and Dictatorship." *American Sociological Review* 5 (1940): 157–69.

Hohenemser, C., M. Deicher, A. Ernst, M. Hofsass, G. Lindner, and E. Recknagel. "An Early Report on the Chernobyl Disaster." *Environment,* June 16, 1986.

Holsti, Ole R. "Crisis, Stress, and Decision Making." *International Social Science* 23 (1971): 53–67.

―――. *Crisis, Escalation, War.* Montreal: McGill-Queens University Press, 1972.

Huffman, James. *Government Liability and Disaster Mitigation.* Lanham, Md.: University Press of America, 1986.

Hewitt, K., and I. Burton. *The Hazardousness of a Place.* Toronto: University of Toronto Press, 1971.

ICFTU and ICEF. *The Trade Union Report on Bhopal.* International Confederation of Free Trade Unions, July 1985.

Janis, I., and L. Mann. *Decision Making.* New York: Free Press, 1977.

Kates, Robert W., ed. *Managing Technological Hazard: Research Needs and Opportunities.* Institute of Behavioral Science, monograph 25. Boulder: University of Colorado Press, 1977.

Kates, Robert W., and Jeanne X. Kasperson. "Comparative Risk Analysis of Technological Hazards." *Proceedings of the National Academy of Sciences* 80 (1983): 7027–38.

Kets de Vries, Manfred F.R., and Danny Miller. *The Neurotic Organization.* San Francisco: Jossey Bass, 1984.

Kinghorn, S. "Corporate Harm: An Analysis of Structure and Process." Paper presented at the *Conference on Critical Perspectives in Organizational Analysis,* Baruch College, The City University of New York, September 4–6, 1985.

Kreps, Gary A. "The Organization of Disaster Response." *International Journal of Mass Emergencies and Disasters* (November 1983): 1.

———. "Sociological Inquiry and Disaster Research." *Annual Review of Sociology* 10 (1984): 309–30.

Kunreuther, H. "Limited Knowledge and Insurance Protection." *Public Policy* 24 (1976): 227–61.

Kunreuther, H., R. Ginsberg, L. Miller, P. Sagi, P. Slovic, B. Borkan, and N. Katz. *Disaster Insurance Protection: Public Policy Lessons.* New York: Wiley, 1978.

Kutak, R.I. "The Sociology of Crisis: The Louisville Flood of 1937." *Social Forces* 17 (1938): 66–72.

La Porte, T. "Public Attitudes Towards Present and Future Technology." *Social Studies of Science* 5 (1975): 373–91.

Lawless, Edward W. *Technology and Social Shock.* New Brunswick, N.J.: Rutgers University Press, 1977.

Leffak, Ellen S. "Product Liability: The Problem of the Non-Designing Manufacturer." *Journal of Products Liability* 9 (1986): 73–87.

Lerbinger, O. *Managing Corporate Crises: Strategies for Executives.* Boston: Barrington Press, 1986.

Levine, A.G. *Love Canal: Science, Politics, and People.* Lexington, Mass.: Lexington Books, 1982.

Lipton, M. *The Urban Bias.* London: Longmaus, 1976.

Liverman, D., and J. Wilson. "The Mississauga Train Derailment and Evacuation, 10–16 November 1979." *The Canadian Geographer* 25 (1981): 365–75.

Lowrance, W. *Of Acceptable Risk, Science, and the Determination of Safety.* Los Altos, Calif.: Kaufman, 1976.

McEachern, A.W. *Organizational Illusions.* Redondo Beach, Calif.: Shale, 1984.

Manual of Industrial Hazard Assessment Techniques. Washington, D.C.: The World Bank, 1985.

Marrone, J. "The Liability Claims Experience of the American Nuclear Pools and their Response to the Three Mile Island." *OECD Symposium on Nuclear Third Party Liability and Insurance,* Geneva, Switzerland, September 1984.

Masuch, Michael. "Vicious Circles in Organizations." *Administrative Science Quarterly* 30 (March 1985): 14–33.

Mazur, A. "Disputes Between Experts." *Minerva* 11 (1973): 55–81.

Meyer, Alan D. "Adapting to Environmental Jolts." *Administrative Science Quarterly* 27 (1982): 515–37.

Meyer, M.W., and K.A. Solomon. "Risk Management in Local Communities." *Policy Studies* 3 (1984): 245–65.

Meyer, Priscilla S., and Jay Gissen. "The Poison Problem." *Forbes,* December 1982, 34–36.

Milburn, T.W. "The Management of Crisis." In *International Crisis: Insights from Behavioral Research*, edited by C.F. Hermann. New York: Free Press, 1972.

Milburn, T.W., R.S. Schuler, and K.H. Watman. "Organizational Crisis. Part II: Strategies and Responses." *Human Relations* 36, no. 12 (1983): 1161–80.

———. "Organizational Crisis. Part 1: Definition and Conceptualization." *Human Relations* 36, no. 12 (1983): 1141–60.

Mileti, Dennis S. "Human Adjustment to the Risk of Environmental Extremes." *Sociology and Social Research* 64 (1980): 328–47.

Mileti, D., J. Sorenson, and W. Bogard. *Evacuation Decision-Making: Process and Uncertainty.* Oak Ridge, Tenn.: Oak Ridge National Laboratory, 1985.

Mitroff, I.I., and R.H. Kilmann. *Corporate Tragedies: Product Tampering, Sabotage and other Disasters.* New York: Praeger Publishers, 1984.

Morehouse, W., and A. Subramaniam. *The Bhopal Tragedy: What Really Happened and What it Means for American Workers and Communities at Risk.* New York: Council on International Public Affairs, 1986.

Mulder, M., J.R. Ritsema van Eck, and R.D. De Jong. "An Organization in Crisis And Non-crisis Situations." *Human Relations* 24 (1971): 19–41.

National Toxicology Program. *Toxicity Testing: Strategies to Determine Needs and Priorities.* Washington, D.C.: National Academy Press, 1984.

Nelkin, D. *Technological Decisions and Democracy.* Beverly Hills, Calif.: Sage Publications, 1977.

Nelkin, D., and M.S. Brown. *Workers at Risk.* Chicago: University of Chicago Press, 1984.

Norris, R., ed. *Pills, Pesticides, and Profits.* Croton-on-Hudson: North River Press, 1982.

Occupational Injuries and Illnesses in 1984. USDL 85-483. Washington, D.C.: Bureau of Labor Statistics, U.S. Department of Labor, November 1985.

Offe, C. *Contradictions of the Welfare State.* Cambridge, Mass.: MIT Press, 1984.

Otway, H.J., et al. "On the Social Aspects of Risk Assessment." *Journal of the Society for Industrial and Applied Mathematics* (1977).

Paige, Glenn D. "Comparative Case Analysis of Crisis Decisions in Korea and Cuba." In *International Crises: Insights from Behavioral Research*, edited by Charles Hermann, 39–55. New York: Free Press, 1972.

Pearson, J. *Technology, Environment, and Development.* Washington, D.C.: World Resources Institute, 1984.

———. *Down to Business.* Study 2. Washington, D.C.: World Resources Institute, 1985.

Perrow, C. *Normal Accidents: Living with High Risk Technologies.* New York: Basic Books, 1984.

Perry, R.W., M.K. Lindell, and M.R. Green. "Crisis Communications: Ethnic Differentials in Interpreting and Acting on Disaster Warnings." *Social Behavior and Personality* 10, no. 1 (1982): 97–104.

———. *Evacuation Planning and Emergency Management.* Lexington, Mass.: D.C. Heath, 1981.

Pfeffer, J., and G.R. Salancik. *The External Control of Organizations.* New York: Harper and Row, 1978.

Pronko, W.H., and W.R. Leith. "Behavior under Stress: A Study of its Disintegration." *Psychological Reports Monograph Supplement* 5 (1956).

Quarantelli, E.L. *Delivery of Emergency Medical Services in Disasters: Assumptions and Realities*. New York: Irvington, 1983.

———. "Chemical Disaster Preparedness at the Local Community Level." *Journal of Hazardous Materials* 8 (1984): 239–49.

———, ed. *Disasters: Theory and Research*. Beverly Hills, Calif.: Sage, 1978.

Randall, W., and S. Solomon. *The Tragedy at Budesberg*. Boston: Little Brown & Co., 1975.

Rayner, Jeannette F. "Studies of Disaster and Other Extreme Situations – an Annotated Selected Bibliography." *Human Organization* 16 (1957): 30–40.

Report of the Inquiry into the Fire at Micheal Colliery, Fire. Command Paper Cmnd. 3657. London: Her Majesty's Stationery Office, 1968.

Report to the Congress: Better Regulation of the Pesticide Exports and Pesticide Residues in Imported Food Is Essential. Washington, D.C.: Comptroller General of the United States, 1979.

Rogers, W., et al. *Report of the Presidential Inquiry Commission on the Space Shuttle Challenger*. Washington, D.C.: U.S. Government Printing Office, 1986.

Rowe, W.E. *An Anatomy of Risk*. New York: Wiley, 1977.

Rubin, I. "Universities in Stress: Decision Making Under Conditions of Reduced Resources." *Social Science Quarterly* 58 (1977): 242–54.

Scapers, R.W., R.J. Ryan, and L. Fletcher. "Explaining Corporate Failure: A Catastrophe Theory Approach." *Journal of Business and Accounting* 8, no. 1 (1981): 1–26.

Schelling, T.C. *Micromotives and Macrobehavior*. New York: Norton, 1978.

Schuler, R.S. "Definition and Conceptualization of Stress in Organizations." *Organizational Behavior and Human Performance* 25 (1980): 184–215.

Schwing, R.C., and W.A. Albers. *Societal Risk Assessment: How Safe Is Safe Enough?* New York: Plenum Press, 1980.

Sea Gem. *Report of the Inquiry into the Causes of the Accident to the Drilling Rig, Sea Gem*. Command Paper Cmnd. 3409. London: Her Majesty's Stationery Office, 1967.

Sethi, S.P. "The Santa Barbara Oil Spill." In *Up Against the Corporate Wall*, 2nd ed., edited by S.P. Sethi, 4–32. Englewood Cliffs, N.J.: Prentice Hall, 1977.

Sheffat, Mary Jane. "Market Share Liability: A New Doctrine of Causation in Product Liability." *Journal of Marketing* 47 (Winter 1983): 35–43.

Shrivastava, P. "Bibliography of Publications on Bhopal." Working paper. New York: Industrial Crisis Institute, December 1985.

———. "Strategic Management of Industrial Crises." Working paper. New York: New York University, April 1986.

———. "A Cultural Analysis of Conflicts in Industrial Disasters." Presented at the XI World Congress of Sociology, New Delhi, August 18–22, 1986.

Sills, D.L., C.P. Wolf, and V.B. Shelanski, eds. *Accident at Three Mile Island: The Human Dimension*. Boulder, Colo.: Westview Press, 1982.

Simons, M. "Some Smell Disaster in Brazilian Industry Zone." *New York Times*, 18 May 1985.

Slovic, P., B. Fischoff, and S. Lichtenstein. "Cognitive Processes and Societal Risk Taking." In *Cognition and Social Behavior*, edited by J.S. Carroll and J.W. Payne. Potomac, Md.: Erlbaum Associates, 1976.

————. "Rating the Risks." *Environment* 21, no. 3 (April 1979): 14–39.

————. "Informing People about Risk." In *Product Labelling and Health Risks*, edited by L. Morris, M. Maris, and I. Barofsky. Banbury Report 6. Cold Spring Harbor, New York: Cold Spring Harbor Laboratory, 1980.

Smart, C.F., and W.T. Stanbury, eds. *Studies in Crisis Management*. Toronto: Butterworth, 1978.

Smart, Carolyne, and Ilan Vertinsky. "Designs for Crisis Decision Units." *Administrative Science Quarterly* 22 (1977): 640–57.

Smets, H. "Compensation for Exceptional Environmental Damage Caused by Industrial Activities." Paper presented at the Conference on Transportation, Storage and Disposal of Hazardous Materials, IIASA, Laxenberg, Austria, July 1–5, 1985.

Sorensen, John H. *Evacuations due to Chemical Accidents: Experience from 1980 to 1984*. Oak Ridge, Tenn.: Oak Ridge National Laboratory, January 1986.

Sorensen, J.H., D.S. Mileti, and E. Copenhaver. "Inter and Intraorganizational Cohesion in Emergencies." *International Journal of Mass Emergencies and Disasters* 3, no. 3 (November 1985): 30–52.

Special Issue on Crisis Management. *The Journal of Business Strategy* (Spring 1986).

Starbuck, William H. "Congealing Oil: Inventing Ideologies to Justify Existing Ideologies Out." *Journal of Management Studies* 18 (1982): 3–27.

Starbuck W.H., A. Greve, and B.L.T. Hedberg. "Responding to Crisis." *Journal of Business Administration* 9 (1978): 111–37.

Starr, C. "Social Benefit versus Technological Risk." *Science*, September 19, 1969, 1232–38.

State of the Environment 1985. Geneva: Organization for Economic Cooperation and Development, 1985.

State of India's Environment: A Citizen's Report. New Delhi: Center for Science and Environment, 1982.

State of the Malaysian Environment Reports. *The Crisis Deepens—A Review of Resource and Environmental Management in Malaysia: 1975–1985*. Malaysia: Sahabat Alam Malaysia, 1986.

Staw, Barry M. "Knee-Deep in the Big Muddy: A Study of Escalating Commitment to a Chosen Course of Action." *Organizational Behavior and Human Performance* 16 (1976): 27–44.

Staw, Barry M., L.E. Sandelands, and J.E. Dutton. "Threat Rigidity Effects in Organizational Behavior: A Multi-level Analysis." *Administrative Science Quarterly* 26 (1981): 501–24.

Suedfeld, P., and P.E. Tetlock. "Integrative Complexity of Communications in International Crisis." *Journal of Conflict Resolution* 21 (1977): 427–41.

Summers, J. "Management by Crisis." *Public Personnel Management* (May–June 1977): 194–200.

Tolba, M.K. *Development Without Destruction: Evolving Environmental Perceptions*. Dublin: Tycooly International Publishing, 1982.

Torry, W.I. "Anthropological Studies in Hazardous Environments: Past Trends and New Horizons." *Current Anthropology* 20 (1979): 517–29.

Turner, B.A. "The Organizational and Interorganizational Development of Disasters." *Administrative Science Quarterly* 21 (1976): 378–97.

————. *Man Made Disasters*. London: Wykeham Publications, 1978.

Tversky, A., and D. Kahneman. "Judgement Under Uncertainty: Heuristics and Biases." *Science*, 1974, 1124–31.

United Nations Center for Human Settlements (HABITAT). *The Residential Circumstances of the Urban Poor in Developing Countries*. New York: Praeger Publishers, 1981.

Volz, William H. "Advising the Wholesaler on Product Liability Exposure." *Journal of Product Liability* 6 (1983): 109–25.

Wallace, A.F.C. *Tornado in Worcester: An Exploratory Study of Individual and Community Behavior in an Extreme Situation*. National Academy of Sciences, National Research Council, disaster study 3. Washington, D.C.: National Academy of Sciences, National Research Council, 1956.

Warheit, J., and J. George. "Fire Departments: Operations During Major Community Emergencies." *American Behavioral Scientist* 13, no. 3 (January–February 1970): 362–68.

Warner, Harland W. "Solid Consumer Relations Can Defuse Crises." *Public Relations Journal* 38 (December, 1982): 8.

Weinstein, N.D. "It Won't Happen to Me: Cognitive and Motivational Sources of Unrealistic Optimism." Unpublished paper, Department of Psychology, Rutgers University, 1979.

Weir, D., and D. Shapiro. *Circle of Poison*. San Francisco: Center for Investigative Reporting, 1981.

WHAZAN. Technica International Ltd. *Computer System for Process Plant Hazard Analysis*. Washington, D.C.: World Bank, 1986.

Wildavsky, Aaron. "Information as an Organizational Problem." *Journal of Management Studies* 20 (1983): 29–40.

Williams, H.B. "Some Functions of Communication in Crisis Behavior." *Human Organization* 16 (1957): 15–19.

Wohlstetter, Roberta. *Pearl Harbor: Warning and Disaster*. Stanford, Calif.: Stanford University Press, 1962.

World Bank. *World Development Report*. New York: Oxford University Press, 1984.

The World Environment Handbook. New York: World Environment Center, 1984.

World Health Organization. *Toxic Oil Syndrome—Mass Food Poisoning in Spain*. Malaysia: World Health Organization, 1986.

World Resources Institute. *Guidelines for Growth: Multinational Corporations and Environment in Developing Countries*. Washington, D.C.: World Resources Institute, 1985.

World Resources Institute. *Improving Environmental Cooperation: The Roles of Multinational Corporations and Developing Countries*. Washington, D.C.: World Resources Institute, 1984.

Zeeman, E.C. "Catastrophe Theory." *Scientific American* 1976, 65–83.

Index

About the Author

Paul Shrivastava, a native of Bhopal, has returned to Bhopal several times since the accident to investigate first-hand its causes and consequences. He conducted more than 200 interviews both in the United States and India, including many Union Carbide officials. Shrivastava is Associate Professor of Management at the Graduate School of Business Administration, New York University. He is also the Founder and Executive Director of the Industrial Crisis Institute, Inc., a nonprofit research organization devoted to analysis of industrial crisis problems. A specialist in strategic management of corporations and administrative problems of developing countries, Shrivastava has spoken on these topics at national and international meetings and has published in numerous professional and scholarly journals. In addition, he is a coeditor of the research annual *Advances in Strategic Management* and a contributing editor to the *Journal of Business Strategy*. He has received numerous awards and grants for his research from the National Science Foundation and from private foundations. Shrivastava holds a bachelor's degree in mechanical engineering, a master's degree in management, and a Ph.D. in management from the Graduate School of Business, University of Pittsburgh.